Territories

Politics and political relationships underpin the world we live in. From the division of the earth's surface into separate states to the placement of 'keep out' signs, territorial strategies to control geographic space can be used to assert, maintain or resist power and as a force for oppression or liberation. Forms of exclusion can be consolidated and reinforced through territorial practices, yet they can also be resisted through similar means. Territoriality can be seen as the spatial expression of power, with borders dividing those inside from those outside.

This extensively revised and updated second edition continues to provide an introduction to theories of territoriality and the outcomes of territorial control and resistance. The book emphasizes the underlying processes associated with territorial strategies and raises important questions relating to place, culture and identity. Key questions emerge concerning geographic space: who is 'allowed' to be in particular spaces? and who is barred, discouraged or excluded? The text explores the construction of territories and the conflicts which often result using a range of examples drawn from various spatial scales and countries. It ranges in coverage from conflicts over national territory, such as Israel/Palestine, Northern Ireland and South Ossetia, to divisions of space based around class, gender and ethnicity. This second edition also contains new contemporary debates on nationalism, territorialization, globalization and borders. It casts light on the territorial consequences of the 'war on terror' and the conflicts in Iraq and Afghanistan. It also examines issues such as migration, the territorial expansion of the European Union and territorial divisions in the home and workplace.

Written from a geographic perspective, *Territories* is an interdisciplinary text drawing upon ideas and material from a range of academic disciplines. The text is richly illustrated throughout with figures, boxed case studies and end of chapter further reading. It will be of interest to undergraduates and graduates studying political geography, politics and international relations.

David Storey is a Senior Lecturer in Geography at the University of Worcester, UK.

Territories
The claiming of space

Second edition

David Storey

Routledge
Taylor & Francis Group

LONDON AND NEW YORK

First published 2001
as *Territories: Nations, States and the Claiming of Space*
by Pearson Education

This edition published 2012
by Routledge
2 Park Square, Milton Park, Abingdon, Oxon OX14 4RN

Simultaneously published in the USA and Canada
by Routledge
711 Third Avenue, New York, NY 10017

Routledge is an imprint of the Taylor & Francis Group, an informa business

British Library Cataloguing in Publication Data
A catalogue record for this book is available from the British Library

Library of Congress Cataloging-in-Publication Data
Storey, David, 1960–
Territories / David Storey. – 2nd ed.
        p. cm.
    Includes bibliographical references and index.
    1. Political geography. 2. Nationalism. 3. Nation-state. 4. Territory,
National. I. Title.
JC319.S76 2011
320.1'2–dc23   2011024555

ISBN: 978–0–415–57549–2 (hbk)
ISBN: 978–0–415–57550–8 (pbk)
ISBN: 978–0–203–85457–0 (ebk)

Typeset in Times New Roman by
Swales & Willis Ltd, Exeter, Devon

MIX
Paper from
responsible sources
FSC
www.fsc.org   FSC® C004839

Printed and bound in Great Britain by
TJ International Ltd, Padstow, Cornwall

Contents

Figures

Acknowledgements

There are many people who have helped (wittingly or unwittingly) in the completion of this book. I have been fortunate to enjoy the support of many colleagues at the University of Worcester; in particular I am grateful to Heather Barrett, Alan Dixon, Nick Evans and John Fagg. Many others including Ronnie Kowalski (sadly missed), James Sidaway, Michael Holmes, Andy Storey and Jennifer Storey have in various ways contributed to this work. Over the years, many students at Worcester have engaged with the issues raised here and have provided useful feedback. The university also granted a period of research leave without which the book would not have been completed. Faye Leerink at Routledge provided much useful advice and assistance throughout and speeded the process up through the sending of numerous email reminders (though these were not always appreciated at the time!). Tamsin Ballard, Richard Willis, Janice Baiton, Lisa Salonen and the production team provided advice and guidance, and helped in constructing the maps. Thanks to you all, and to the many others not mentioned here. Anne undertook the practical task of proofreading and critically commenting on the manuscript but, more importantly, I am indebted to her for dealing cheerfully with my whims and downright foolishness, enduring my mood swings and obsessions, and for displaying remarkable understanding and patience.

1 Introduction

We live in a highly territorialized world where we are regularly confronted with signs such as 'authorized personnel only', 'no trespassing', 'prohibido el paso', and so on. Such everyday warnings and admonishments are a reflection of attempts to impose forms of power (through rules and regulations) over portions of geographic space (Figure 1.1). They are manifestations of the intersections between geography and politics and are highly visible reminders of the ways in which power is imposed, accepted or resisted. Political geography draws attention to the spatial dimensions of power, dealing with political phenomena and relationships at a range of spatial scales from the global down to the local. The sub-discipline revolves around the intersections of key geographical concerns of space, place and territory on the one hand and issues of politics, power and policy on the other.

A glance at the global political map provides us with the most obvious manifestation of the territorial dimension to politics in the division of the earth into separate countries or states. However, this macro-scale territorialization is accompanied by a myriad of much more micro-scale variants including those mentioned above. In everyday usage, territory is usually taken to refer to a portion of geographic space which is claimed or occupied by a person or group of persons or by an institution. Following from this, the process whereby individuals or groups lay claim to such territory can be referred to as 'territoriality'. However, as will be seen, these somewhat simplified definitions mask considerable complexity. The investigation of various dimensions of territorial formation and behaviour at different spatial scales forms the central focus of this book. It considers the ways in which territories are 'produced' and explores how territoriality is used as a strategy to assert power or to resist the power of others. It examines the meanings of territory, the construction of territories and the use of territorial strategies at the level of the state, sub-state divisions, at the micro-scale level in individual localities, in the workplace and

Figure 1.1 'No Entry' sign

within the home. The next three sections of this chapter provide a brief introduction to these themes and the chapter concludes by providing an outline of the book's structure.

Territory at a macro-scale

At a global level, the pursuit of spheres of influence by major powers, whether in the eras of formal colonialism, superpower tension, or other contemporary forms of geopolitical rivalry, represents a distinct version of territorial behaviour. In Europe, following the demise of feudalism, older political entities were replaced initially by a series of city-states and larger territorially based units. In fact, the origins of the word *territory* can be traced back to medieval times. In the Roman era, the word *territorium* was associated with both community and territory. Slowly, the idea of owing allegiance to the territory began to supersede allegiance to a lord, or to God. Wars were to be fought in the name of territorial formations. This is not to suggest some significant difference in the underlying purpose. Wars are generally centred on issues of power and control but the significance arose from the fact that now they were being fought utilizing territorial terminology. Territory had become the mobilizing force and, to a considerable extent, control of territory became the geographical expression of political power.

Ultimately, the present inter-state system evolved into a more cohesive territorial political system in which almost all the world's territory is seen to belong to distinct political entities commonly known as states (though some of these territories are subject to competing claims). This process was facilitated with the advent of colonialism, a political project whereby a number of European countries engaged in a process of territorial accretion beyond the confines of Europe through the process of acquiring colonies. Empire-building by European powers, most notably Spain, Portugal, The Netherlands, France and England, represented a territorial expression of power on the part of these countries which led to much of the world's surface area being carved up between them (Figure 1.2). While the bulk of these colonies have since attained independence, there has been an enduring territorial legacy (together with a range of economic, political and cultural consequences) throughout Africa, Asia and the Americas.

More recently, the Cold War era of the middle years of the twentieth century saw the United States and the Soviet Union establishing, or seeking to establish, geographical spheres of influence. This was most notable in the Soviet Union's effective control of eastern Europe and the USA's flexing of its political and military muscle in its own 'backyard' of central America. While this might not have been formalized territorial control, it was a process whereby 'friendly' governments were encouraged, or sometimes forcibly installed, in countries which the superpowers saw as vital to their strategic interests. In this way, the world was effectively divided between those who were allied to one side or the other, with a small group of neutral countries (Figure 1.3). Within the contemporary geopolitical context, the USA can be seen as the pre-eminent military superpower and, in the wake of the 11 September 2001 attacks on its territory, it has led invasions of Afghanistan and Iraq. Together with its string of military bases in various parts of the world, it asserts something of a policing role which for many is akin to a contemporary version of colonialism through which subject peoples and territories are maintained in a subordinate position (Harvey 2003; Gregory 2004). While the presence of US bases does not correspond to outright territorial control, it does allow for the exertion of considerable political and military influence (Figure 1.4).

The emergence of regional economic alliances is another form of territorialization, epitomized by the evolution of trading blocs such as the European Union (EU) and the Association of South East Asian Nations (ASEAN). The current trajectory of the EU, from a trading bloc, through an economic union and towards greater political integration, has been allied to a policy of continued spatial expansion. To some, this type of supra-state confederation is seen as the successor to the state system in an increasingly globalized world where state boundaries are assumed to have diminishing significance. Below this global level in the spatial hierarchy is the system of states already referred to, the most

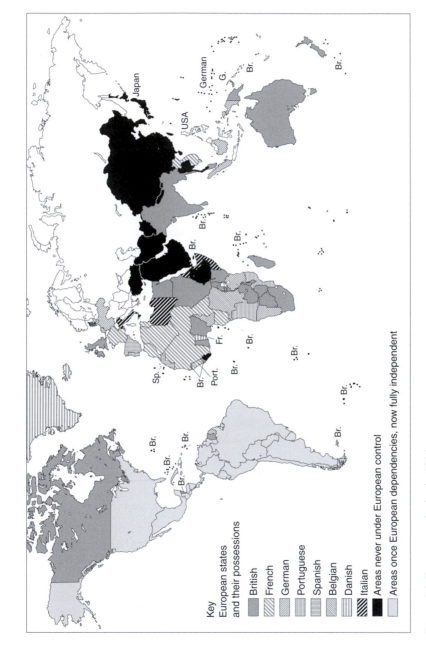

Key
European states
and their possessions

British
French
German
Portuguese
Spanish
Belgian
Danish
Italian
Areas never under European control
Areas once European dependencies, now fully independent

Figure 1.2 European colonies in 1914

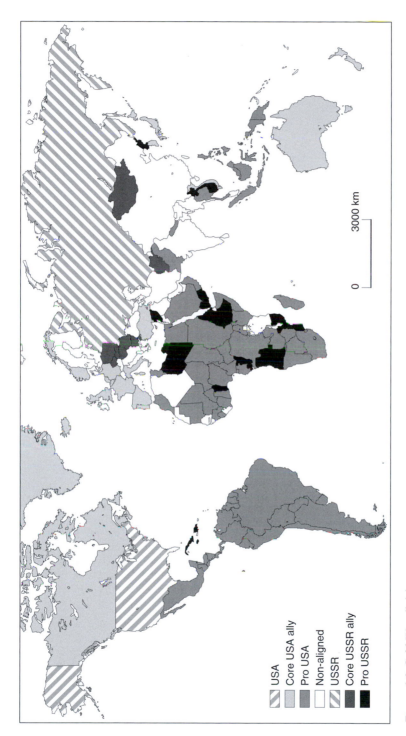

Figure 1.3 Cold War divisions

USA
Core USA ally
Pro USA
Non-aligned
USSR
Core USSR ally
Pro USSR

3000 km

0

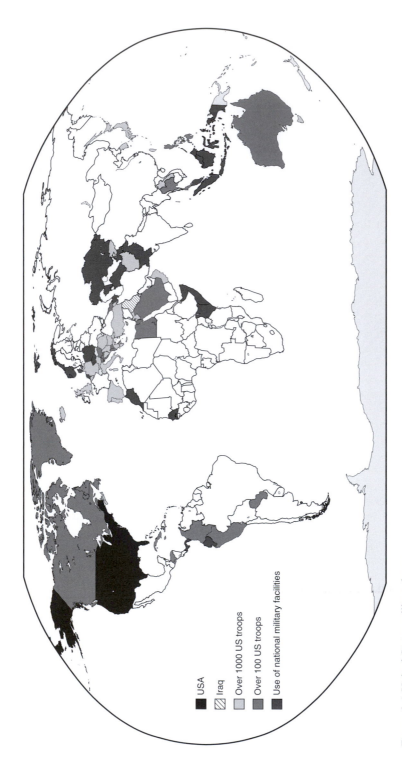

Figure 1.4 United States military bases

obvious example of formalized territorial organization in the world today. The political division of the planet into territorial states is a well-known and 'taken for granted' feature. There is a general acceptance of the idea of formalized and rigidly demarcated boundary lines even though the existence of some of these, and the precise location of others, may be subject to dispute in certain instances. The process of state formation has been facilitated through the concept of the nation and its associated powerful territorial ideology of nationalism.

Territoriality can also be observed at the sub-state level. The internal divisions of a country, whether through local or federal government, represent a formalized internal territorialization. German *Lander*, English counties or the individual states of the USA are all sub-state territorial formations which are integral parts of the larger polity. Neo-liberal economic philosophies in recent decades have emphasized a need to 'roll back the state' and reduce its role as a provider of public utilities. While this has created spaces for private companies to provide a range of services, it has also been accompanied by a greater emphasis on community-based (often voluntary) responses to social and economic concerns. This has led to the growing importance of both quasi-governmental institutions and locality-based organizations. These local responses to issues represent another version of territorialized political behaviour, whereby spatially defined communities seek solutions to problems within their own areas.

Territory at a micro-scale

The instances mentioned so far are the more obvious examples of territorial formations and might also be seen as the traditional core of the sub-discipline of political geography. While the state and its associated geographical characteristics, most notably territory and borders, have been a key focus, the last few decades of the twentieth century saw a reinvigoration of the sub-discipline through the introduction of more radical and politically engaged perspectives. More recently still, elements of social theory have been incorporated, both deepening and broadening its conceptual base. There is now a much more critical approach and a more diverse one being brought to bear on a subject matter that extends well beyond the realms of the state. While much contemporary political geography maintains a focus on what might be seen as 'big' politics (states, governments, etc.), there has been an increased concern with 'small' politics (local issues, gender, ethnicity, social identities) (Cox 2002; Jones *et al*. 2004; Blacksell 2006; Painter and Jeffrey 2009; Storey 2009). This suggests a need to consider territory as any (even informally) bounded space and that such spaces may experience internal or external contestation. For example, within urban areas, clear spatial divisions can be seen to exist with fault lines dividing zones in terms of affluence, class or ethnicity. As a result, particular areas become characterized in some discourses as, for example, 'poor', 'working class', or 'Asian'. Such divisions of urban space into informal territories lead to a consideration of the ways in which space can be seen as gendered or racialized. For example, residential concentrations of particular ethnic groups, a clear territorial feature, occurs in many urban areas. These often come to be seen as the territory of the particular ethnic group concentrated there. Similarly, other spaces may be divided in terms of gender. For example, football pitches (and indeed football grounds generally, not just the pitch) have traditionally been seen as male spaces while the kitchen has been seen as a female space.

Owning a house can also be seen in territorial terms. Private property is regarded by many as a logical and necessary outcome of human territorial behaviour. It represents a claim to space reinforced by the legal systems of many countries. Taking the idea of territory down to its most elementary level, the desire for personal space can be seen as a form of territorial behaviour. Humans like to have a pocket of space around them which is 'theirs' and they resent others 'invading' their space. This can be interpreted as a territorial claim to a portion of geographic space. These examples also lead to a consideration of the concepts of public and private space and the divisions between them.

Territory and power

Contemporary concerns within geography extend well beyond mere descriptions of territorial formation and behaviour. There is a need to explain why such behaviour occurs. It is the central premise of this book that territorial strategies are employed by individuals or groups in order to attain or to maintain control. Whether explicit or implicit, control over territory is a key political motivating force and the apportioning of space or specified territory results from the interplay of social and political forces. Territoriality can be seen as the spatial expression of power and the processes of control and contestation over portions of geographic space are central concerns of political geography (Cox 2002; Paasi 2008; Delaney 2005, 2009; Dahlman 2009). Any consideration of territories necessarily raises questions to do with boundaries. The borders of a territory may be quite clearly defined and formalized, as between most countries, or they may be somewhat informal or ill-defined, as is the case when considering the territories of rival urban gangs whose 'patch' or 'turf' may well have a core but the precise boundaries of which may be unclear. Territoriality and the construction, imposition and maintenance of boundaries are political strategies designed to attain particular ends. From all of this it is obvious that territories are often the centre of disputes (in many instances quite violent) and the areal extent of particular territories (and even their right to exist) may be regularly contested.

Contemporary debates in cultural geography use ideas of social and spatial exclusion and notions of transgression in highlighting unequal access to particular spaces. The ways in which 'others', or non-dominant groups, are seen to be excluded while these groups, in turn, resist those processes which actively exclude them are an important focus. As will be seen in the examples which follow, non-dominant groups often attempt to wrest control of geographical space in order to assert their presence. People resist power and the imposed territorial boundaries through which that power is expressed. The borders erected around territories, whether formal or informal, are subject to periodic contestation and transgression. In practical terms, it may make a considerable difference which side of the line a person finds themselves on. Hence, the desire by many Mexicans to cross the Rio Grande into the United States or the risks taken by refugees to attain a safe haven outside their own country. Such strategies are set against the background of a 'postmodern' world in which, some argue, boundaries are becoming more and more permeable and ephemeral as processes of globalization increase in intensity and formal state borders are seen as being of less and less importance. Linked to this, rapid advances in information and communications technology have intensified and speeded up global transactions leading to suggestions that cyberspace developments are making borders redundant. While such claims of a borderless world seem wide of the mark, territories are not static and are constantly produced and re-produced. Although efforts are continuously made to reinforce territorial control, these are often resisted and space is regularly transgressed.

While highlighting the functions of territory, the phenomenon should not be reduced to a mere commodity or be seen simply as a spatial container. As will become clear, territories, and the ways in which they are imagined, can play an important role in the formation of people's self-identity. The chapters which follow draw attention to the ways in which territorial identities are important components in our overall sense of identity. This book explores the theoretical underpinnings lying behind different expressions of territoriality. This theoretical material is supported by a variety of historical and contemporary examples ranging from consideration of overt political conflicts, such as those surrounding the north of Ireland and Israel/Palestine, to issues such as local government and domestic space. While it is obvious that the processes giving rise to national political conflicts are very different from those associated with, for example, divisions of space in the domestic sphere, nevertheless, they pertain to human attempts both to control portions of space and to resist that control. As such, the many diverse types of territory and forms of territorial behaviour considered in this book are connected in the sense of demonstrating people's interactions with place and space. Furthermore all are intimately bound up with issues of power.

While the structure of this book may appear to imply a sharp distinction between the macro and micro scales, it should be obvious that this division is somewhat artificial. Divisions between the local, national and global are of course far from discrete and the manner in which scale is 'constructed' has engendered much debate (Brenner 1998; Cox 1998; Herod and Wright 2002). Many macro-level processes, such as state-building, have quite localized effects while events in particular places may well impact at more global levels. The macro and the micro are not structurally separate geographic spaces. This book is concerned with the processes through which territories are produced and imagined, the contested nature of territorial formation and control, the use of territorial strategies, and the importance of territorial identity. These form the key themes which underpin the book. As a result, ideas of power, dominance, resistance, conflict and identity run throughout each chapter.

Structure of the book

Chapter 2 explores the concept of territory and provides an overview of theories explaining human territoriality. The direct political control over a designated territory and the processes of state formation are the subject of Chapter 3. Associated with state formation is the territorial ideology of nationalism. Concepts of nation and nationalism, and the symbolism associated with them, form the subject matter of Chapters 4 and 5. Chapter 6 focuses on debates surrounding the future of the state as a sovereign territorial entity, particularly in light of arguments suggesting that borders are of declining significance. Territoriality at the sub-state level, as evidenced through internal sub-divisions and the promotion of community and locality groupings, is dealt with in Chapter 7. The more micro-scale versions of territoriality – those which occur at relatively local levels, such as the 'racialization' of space by particular ethnic groups (whether dominant or subordinate) or the informal claiming of space on the basis of gender – are explored in Chapter 8. Chapter 9 provides a summary of the key points. In focusing on formal and informal territorial strategies, and in emphasizing the contested nature of territory, this book deals with aspects of what might be seen as the traditional subject matter of political geography, namely a concern with the state and the nation. However, in exploring other dimensions of territorial control and focusing on various spatial strategies of dominance and resistance, this book incorporates more recent developments in human geography and places these within a consideration of power relationships viewed from a spatial perspective.

Further reading

Suggested reading on the issues dealt with in this book is provided at the end of each chapter. The list below contains some general introductory geography texts and academic overviews of the changing nature of geographical thought and practice which can usefully be consulted alongside this volume. The list also contains wide-ranging political geography texts and collections of key writings.

Agnew, J. (2002) *Geopolitics. Revisioning World Politics*, 2nd edn, London: Routledge.

Agnew, J. Mitchell, K, and Toal, G. (eds) (2008) *A Companion to Political Geography*, Oxford: Blackwell.

Blacksell, M. (2006) *Political Geography*, London: Routledge.

Cloke, P., Crang, P. and Goodwin, M. (eds) (2005) *Introducing Human Geographies*, 2nd edn, London: Hodder Arnold.

Cox, K. R., Low, M. and Robinson, J. (eds) (2008) *The Sage Handbook of Political Geography*, London, Sage.

Daniels, P., Bradshaw, M., Shaw, D. and Sidaway, J. (eds) (2012) *An Introduction to Human Geography. Issues for the 21st Century,* 4th edn, Pearson: Harlow.

Flint, C. and Taylor, P. (2007) *Political Geography World-Economy, Nation-State and Locality*, 5th edn, Harlow: Pearson.

Gallaher, C., Dahlman, C. T., Gilmartin, M., Mountz, A. (2009) *Key Concepts in Political Geography*, London: Sage.

Johnston, R. and Sidaway, J. (2004) *Geography and Geographers. Anglo-American Human Geography since* 1945, 4th edn, London: Arnold.

Jones, M., Jones, R. and Woods, M. (2004) *An Introduction to Political Geography: Space, Place and Politics*, London: Routledge.

Livingstone, D. (1992) *The Geographical Tradition. Episodes in the History of a Contested* Enterprise, Oxford: Blackwell.

Ó Tuathail, G. (1996) *Critical Geopolitics. The Politics of Writing Global* Space, London: Routledge.

Ó Tuathail, G., Dalby, S. and Routledge, P. (eds) (2006) *The Geopolitics* Reader, 2nd edn, London: Routledge.

Painter, J. and Jeffrey, A. (2009) *Political Geography*, 2nd edn, London: Sage.

Peet, R. (1998) *Modern Geographical* Thought, Oxford: Blackwell.

Phillips, M. (ed.) (2005) *Contested Worlds. An Introduction to Human Geography*, Aldershot: Ashgate.

Political Geography (2010) 29 (5) Special Issue on the State of Critical Geopolitics.

2 Territory and territoriality

As indicated in Chapter 1, there is an observable tendency for humans to engage, either individually or collectively, in forms of territorial behaviour. We lay claim to geographic space on an everyday basis. This chapter introduces ways of conceptualizing territory and territoriality in human societies. An initial starting point is to consider whether territorial behaviour should be viewed as an intrinsic part of human nature. The manifestations of human territoriality lead people to sometimes assume that humans have a natural tendency to behave in a territorial manner: to claim space and to prevent others from encroaching on 'our' territory. Territorial behaviour is bound up with staking claims to geographic space and with the production of territories. The concept of territory itself has received less attention than territoriality but, following a consideration of territorial behaviour, the idea of territory is explored later in the chapter. Finally the links between territorial formation and behaviour, and senses of identity and belonging, are examined.

Considerable debate has occurred over the extent to which territorialization and territorial behaviour should be seen as 'natural' or 'social' phenomena, a debate echoing wider long-standing arguments over the relative influence of nature and nurture, a division that many see as somewhat artificial and itself a discursive construction (Whatmore 2002). Some biological and genetic theories (or popular interpretations of them) argue that territoriality is an innate feature of all species, including humans, and that forms of territorial behaviour can, therefore, be seen as natural. Juxtaposed to these deterministic ideas is a body of thought which rejects this naturalizing of territoriality and suggests that territorial behaviour in humans is a phenomenon arising out of our broader socio-political conditioning. Each of these broad sets of theories is outlined in the sections which follow. It is important to bear in mind that these theoretical overviews are presented in somewhat simplified forms in order to convey their essence. It should also be borne in mind that there is a variety of theoretical positions which exist somewhere on a spectrum ranging from totally biological theories through to totally 'conditioned' theories. It might be argued that an understanding of territorial behaviour requires a theory situated somewhere in between these polar positions.

Naturalizing territoriality

Traditionally, the discussion of territoriality has been led by biologists, anthropologists and psychologists. Given their biological basis, it is not surprising that many of these arguments tend to support the idea of territorial behaviour as 'natural' rather than learned or contrived. Essentially two key positions emerge out of this. First, there is a deterministic perspective which sees the acquisition of territory as a natural phenomenon. The second more nuanced perspective sees territorial behaviour in a behaviourist but non-deterministic sense. In its crudest form, the deterministic argument holds that the need for space is a characteristic innate to all species, including humans. When allied to theories suggesting that aggression is a natural or instinctive phenomenon (Lorenz 1966), this desire for space leads 'naturally' to the acquisition of territory, by the use of aggressive behaviour if necessary. There is, then, an impulse to defend this territory against others seeking to 'invade' it.

Utilizing ethology (the study of animals in their natural environment), a set of arguments has been developed whereby human behaviour is seen to mirror that of animals. These views have been widely disseminated through the writings of people such as the anthropologist Robert Ardrey, who argued that animals and humans behave in an intrinsically territorial manner. Humans have what Ardrey (1967) describes as a 'territorial imperative' which compels them to defend space. This is a crude biologically determinist position from which territorial behaviour in humans is seen as a natural and unchanging phenomenon. Parallels between the behaviour of humans and animals were popularized by the zoologist Desmond Morris in his many books, such as *Manwatching* (1973) and *The Naked Ape* (1994). Morris argued that humans are simply another species of animal and, as such, human behaviour patterns are primarily a result of genetic programming. One of these behavioural traits is the defence of territory. Humans are, in the view of Morris, territorial creatures. Relying heavily on the work of Lorenz, Morris argues that we have an innate need to defend territory, whether at the level of the nation or at the micro-scale level of our daily habitat, in the home. Morris sees our tendency to adorn our homes in particular ways as a means of asserting our individuality through placing our 'stamp' on our own territory. In this way, the simple act of painting the front door can be read as an assertion of 'territorial uniqueness'. The socio-biologist Richard Dawkins, in his popular book *The Selfish Gene* (1976), argued that humans act as mere containers for our genes. From this, it is argued that people are genetically programmed to defend those who are like themselves. This leads to territorial defence against people who are different. While most scientists would not endorse such implications, some interpretations of this are used to bolster views that violent conflict and phenomena such as selfishness, racism and xenophobia are natural and inevitable.

These analyses of territorial behaviour, it has been argued, hold true both at the level of the individual defending her or his property and at the collective level of the state defending itself against its rivals. One geographer who leaned heavily towards biological theories was Friedrich Ratzel. In Germany, in the latter part of the nineteenth century, he developed organic theories of state formation which will be returned to in Chapter 3. Suffice to say for the moment that Ratzel's thinking led him to view state expansion as a necessary means of ensuring a country's survival. He borrowed from Darwin's evolutionary theory in suggesting that states might need to adopt a 'survival of the fittest' strategy in order to retain power. This serves to justify aggressive strategies of territorial defence and acquisition. More recently, the geographer Malmberg, from a biologically determinist position, asserted that 'territorial defence is based on this instinctive aggression' (1980: 26). A more nuanced argument suggesting an innate basis for territorial behaviour derives from the work of the psychologist Piaget who examined children's need to feel secure, emphasizing the importance attached to familiar surroundings (Piaget and Inhelder 1967). Consequently, it is argued that children need space in which they feel comfortable and safe and which they can regard as their own. Assuming this requirement stays with us as adults, it follows that there is a need to claim our space and, hence, to assert and maintain some form of control over it, either directly or indirectly.

The arguments outlined above form a component in the wider debate between innate and learned behaviour or between nature and nurture. Determinist arguments attribute many human characteristics to nature and employ ideas from the natural sciences in an attempt to explain human behaviour. From this perspective, territorial behaviour is viewed as being innate to humans rather than being a product of our social and cultural conditioning. These arguments have a certain appeal in that they appear to provide an explanation for elements of observable human behaviour, whether related to international conflicts or the activities of rival football fans. In essence, the argument is that territorial behaviour is a consequence of our evolutionary past rather than our cultural present.

While not rejecting totally the parallels between the behaviour of humans and other species, there are a number of criticisms which can be made. First, the arguments used by Ardrey and others could be seen as bad science. There are three main points here. The first is that, in many instances, there is a distinct lack of empirical evidence to support the suggestions being made. This renders the conclusions more akin to assertions rather than to meaningful interpretations. The second point is that the examples used are highly selective. Ardrey uses examples which support his case. He might have used others which would cast grave doubts on his assertions. The third 'scientific' weakness is the huge leaps in logic employed by Ardrey and others. He assumes that human behaviour can be extrapolated from the behaviour of certain animals, more a leap of faith than a scientifically proven fact. It has been argued that such extrapolation is questionable due to differences in cognitive abilities between people and animals, humans' capacity for language and culture, and the sheer diversity of animal species and human behaviour (Hinde 1987). The extent to which it is valid to infer group behaviour on the basis of a small number of individual occurrences is also open to question. In short, Ardrey tends to assert rather than prove the existence of a territorial imperative.

A second type of criticism relates to apparent contradictions in some of the arguments, or at least to the fact that the evidence may be open to alternative interpretations. To amass evidence of territorial behaviour in humans is not in itself proof that we are innately territorial. It might just as easily be seen as proof that we are all conditioned in broadly similar ways. In rejecting overtly biologically determinist theories, the anthropologist Alland argues that human nature is more open (Alland 1972: 24). He suggests that there is a false dichotomy between biological and social conditioning; rather, he argues, the two interact so that humans, instead of being programmed to respond in particular ways, can decide on a course of action dependent on the opportunities available to them. Similarly, the evolutionary biologist Stephen Jay Gould (1983, 1991) rejected the simple binary opposition between nature and nurture and argued that while biology is hugely influential, it presents a series of opportunities rather than setting limits to our behaviour. For Gould and for others, traits such as kindness and peacefulness might be regarded as equally biological as others such as aggression and violence; our behaviours are not pre-determined. The fact that different cultures have different modes of behaviour leads to the suggestion that the supposed innate basis of human territoriality is less strong than Ardrey, Morris and others would have it. Morris (1973) talks of the human need for personal space but in doing so he points to the fact that the amount of space needed varies from society to society, a fact noted long ago by Hall (1959). This suggests that social or cultural conditioning (the nurture side of the nature–nurture debate) plays a key role in explaining territorial behaviour.

A third strand of criticism relates to the obvious ideological underpinnings of the arguments used by some, most notably Ardrey. Crude biologically based arguments often (wittingly or unwittingly) lend support to particular ideological positions (Rose *et al.* 1990). The arguments of Ardrey and Morris, for example, tend to justify aggression and 'naturalize' conflict. Animosity towards particular ethnic groupings or towards others recognizably 'different' may then be excused, thereby justifying racist or sexist behaviour. In addition, in presenting territorial defence as a natural phenomenon, they provide useful support for the institution of private property. Humans' desire to own and possess is seen as natural rather than a cultural product. Thus, Ardrey criticizes collective farms in the then Soviet Union on the grounds that their inefficiency derived from the fact that people's natural connection or affiliation with territory has been broken. In this way, private property becomes naturalized; it is seen as a consequence of our genes rather than as a function of political relationships and a biological justification is used to support an ideological position. The tension here is between a view of 'nature' used to support a culture of individuality and competition, as opposed to one of mutuality and co-operation. Gould has referred to biological determinism as a 'sociopolitical doctrine masquerading as science' (1991: 302).

From the above criticisms, it follows that more nuanced explanations are required. A number of social psychological and ethological theories have been advanced which suggest that human and animal behaviour might be seen as analogous rather than homologous, as Ardrey and others suggest. In other words, there may well be behavioural parallels between animals and humans but this is not proof that the motivating factors are the same. Many social and environmental psychologists would argue that behavioural similarities do not necessarily reflect similar processes:

> territoriality represents a culturally derived and transmitted answer to particular human problems, not the blind operation of instinct . . . Its rules, mechanisms, and symbols are developed gradually over time and are passed from one generation to the next by the . . . process of socialization.
>
> (Gold 1982: 48)

As suggested earlier, more recent work has pointed towards the need to examine the balance between evolutionary and social forces in shaping our behaviour. Human behaviour occurs within a cultural context and alternative more co-operative traits may be just as 'natural' as aggression and competitiveness (McCullough 2008; Verbeek 2008).

While most scientists would not necessarily endorse the ideological implications of the biologically determinist positions alluded to above, some elements of this line of thought retain a degree of popular credence so that certain forms of behaviour tend to be excused as 'natural', rather than being related to our social environment or wider cultural circumstances. It appears obvious that alternative explanations of territorial behaviour need to be explored. There is a need to move beyond ideas based primarily on genetics to ones which consider the social context in which humans exist.

Socialized territoriality

There is a major strand of thinking in the social sciences which argues that human behaviour, far from being innate, is heavily conditioned by the wider economic, social, political and cultural environment in which we exist. It is suggested that much of human behaviour is learned and not natural. This emphasis on social conditioning leads to a consideration of socio-political influences. With respect to territoriality, a body of thought exists which rejects the determinism of many crude psychological and ethological arguments and which suggests that human territorial behaviour is a product of conditioning rather than a biological urge. Central to this is an emphasis on power relationships rather than biology.

Two geographers who have contributed significantly to this debate are Jean Gottmann and Robert Sack. Gottmann observed that people 'have always partitioned the space around them carefully to set themselves apart from their neighbours' (1973: 1). He highlighted the fact that territory, as he saw it, represented a portion of geographical space under the jurisdiction of certain people. Territoriality reflected the relationship between people and place and its wider significance arises from the fact that 'it signifies also a distinction, indeed a separation, from adjacent territories that are under different jurisdictions' (1973: 5). Thus, territory and the assertion of control over it, represents an expression of power in which the clear message is 'we control this space'.

Gottman identified two reasons for territoriality. First, it confers security. Territory can be converted into defensible space. Second, he argued that it can provide opportunities through facilitating the economic organization of space. This allows people to pursue 'the good life'. In this, he was referring back to both Plato and Aristotle. Plato was concerned with the creation of territorial units which would be self-sufficient and secure. There was also a moral dimension to Plato's thought in that he felt that moral goodness could be attained or maintained through this territorial system, idealized in the city-states or *polis* of medieval Europe. Aristotle, on the other hand, argued that territorial units should not isolate themselves but should actively participate in trade. In this way, the formalizing of a territorial system offered opportunities for advancement and not just security. This consideration of the issues of security and opportunity suggests a political and economic basis for territorial formation. Territorial states, with their systems of administration and centralized control, allow for a standardization of currency and economic regulations and may be seen as a more efficient means of political organization than systems of overlapping jurisdictions.

In his landmark book *Human Territoriality: Its Theory and History*, published in 1986, Robert Sack cautions against determinist views of human territoriality as a basic instinct and emphasizes instead the role of territoriality as a geographic and political strategy (Box 2.1). For Sack, territoriality is 'the attempt by an individual or group to affect, influence, or control people, phenomena, and relationships, by delimiting and asserting control over a geographic area' (1986: 19). He draws attention to the means through which territorial strategies may be used to achieve particular ends. In essence, the control of geographic space can be used to assert or to maintain power, or, importantly, to resist the power of a dominant group.

> Territoriality, as a component of power, is not only a means of creating and maintaining order, but is a device to create and maintain much of the geographic context through which we experience the world and give it meaning.
>
> (Sack 1986: 219)

Territoriality is deeply embedded in social relations, and territories emerge as a consequence of social practices and processes rather than being natural entities (Delaney 2005).

Box 2.1 Robert Sack on territoriality

Surprisingly, given geography's preoccupation with space and place, Robert Sack is one of the few geographers to have attempted a serious analysis of the concept of human territoriality. His book *Human Territoriality. Its Theory and History* was published in 1986 and follows on from an earlier paper (Sack 1983). In it, Sack rejects determinist theories of territoriality in favour of a 'political' theory which sees territorial behaviour as a geographic strategy rather than a basic instinct. Space and society are inter-connected and territoriality is the process which connects them. Sack sees territories as social constructs rather than natural phenomena. Using both historical and contemporary examples, he argues that territoriality is embedded in social relations. He uses examples drawn from the territorialization of North America, strategies employed by the church and the partitioning of space in the home and the workplace to highlight the role of territoriality as a component of power. He concludes that it is 'a device to create and maintain much of the geographic context through which we experience the world and give it meaning' (1986: 219). Territories are human creations, produced under particular circumstances and designed to serve specific ends. Once these territories have been produced, they become the spatial containers within which people are socialized.

Sack's work is important in not only emphasizing the political context of territorial behaviour but also in highlighting how territoriality as a strategy operates at all spatial scales from the geopolitical strategies of superpowers down to the home and the workplace. More significantly, Sack identifies a number of what he terms 'tendencies of territoriality'. Chief among these is that territoriality involves a classification by geographic area. Space can be apportioned between states or between individuals – this room/office/desk is *mine*, not *yours*. However, territories are more than mere spatial containers; they link space and society and convey clear meanings relating to authority, power and rights (Sassen 2006; Delaney 2009). Second, territoriality is easy to communicate via the use of boundaries. These boundaries indicate territorial control and, hence, power over prescribed space. A territory is 'a bounded social space that inscribes a certain social meaning onto defined segments of the material world. A simple territory marks a differentiation between an "inside" and an "outside"' (Delaney 2005: 14). These geographic spaces convey messages of political power and control which are communicated through various means, most notably through the creation and maintenance of *boundaries*. These and other mechanisms facilitate control over space and of those within it; they serve as divisions between those who are inside and those who are outside. In this way, territory allows classification, communication and enforcement through boundary-making. The meanings conveyed may have important implications such as constraining, restricting or limiting mobility for example. A third tendency of territoriality identified by Sack is that it functions as a means through which power is reified. Through the visibility of land (or a room or a desk) power can be 'seen'. Perhaps most importantly, through this, reification of territoriality attention is deflected from the power relationship so that 'territory appears as the agent doing the controlling' (Sack 1986: 33). As an encapsulation of this, Sack notes the use of the term 'the law of the land' whereby we are instructed to obey laws enacted in the name of a territorial formation rather than a set of people in power. Territories can thus be seen as useful political devices that, taken together with territorial strategies, tend to reify power so that it appears to reside in the territory itself rather than in those who control it. Attention is thereby deflected away from the power relationships, ideologies and processes underpinning the maintenance of territories and their boundaries. In this way, 'territory does much of our thinking for us and closes off or obscures questions of power and meaning, ideology and legitimacy, authority and obligation' (Delaney 2005: 18). Ultimately territoriality can be seen as 'a primary geographic expression of social power' (Sack 1986: 5).

One of the problems of assessing the relative merits of the biological and social conditioning arguments is that conclusive proof one way or the other is virtually impossible. Separating out those elements of human behaviour which may reflect innate characteristics from those which are conditioned is not possible. Soja suggests that 'aggression and territoriality may be characteristic of the human species but are not inevitable, ineradicable, or invariable' (1971: 31). He argues that, although there are similarities between human and animal behaviour, the motivation may be substantially different. Equally, Alland (1972) argues that territoriality, in facilitating an efficient use of resources (though clearly this is not always the case), may allow for co-operation rather than leading to outbursts of aggressive behaviour. Beyond the more populist versions of biological determinism referred to above, more nuanced academic socio-biological and evolutionary psychology theories point to the interplay of genetic inheritance and cultural environments in shaping human behaviour (see Workman and Reader 2008). Rather than adopting simplistic deterministic explanations, it would seem reasonable to conclude that we are subject to a complex set of influences within our broader environment and these influences interact with our genetic potential. Rather than getting bogged down in a nature versus nurture debate we might wish to consider the range of options available to us as humans to shape the world we are in, rather than simply assuming we are programmed to behave in particular ways. Our world can, to some extent at least, be shaped in a range of possible ways, rather than being determined in one of a limited number of directions.

The production of territory

While Sack's work focused usefully on territoriality as a political strategy, more recently Stuart Elden has drawn attention to the concept of territory itself (2005, 2010). He suggests that the term has been used somewhat uncritically but he argues that the manner in which the concept emerged is bound up with particular ways of thinking about geographic space; ways which reflect notions of power and control (Box 2.2). From this, it follows that it is not just territorial strategies that need to be interrogated, but the broader ways in which we conceive of space. Elden (2005) contends that developments in cartographic techniques and advancements in geometry and mathematical calculation led to our modern conceptualization of space. Territories are, therefore, more than simply land; instead, territory can be viewed as a political technology, related to the measurement of land and the control of terrain. As such, it is suggested that both territories themselves and their boundaries (concepts which we tend to take for granted) reflect a distinctive mode of social and spatial organization which is dependent on particular ways of calculating and thinking about space. The features of territoriality identified by Sack (1986), in allowing classification and differentiation, are thus a product of the way in which space is imagined and territories are in effect politicized space: mapped and claimed, ordered and bordered, measured and demarcated (Elden 2007a; Brighenti 2010a). This takes us back to Gottman's (1973) view of territorial formation as a more efficient means of administering and organizing space. In a similar vein, Painter (2010) argues that territory is an effect or an outcome of a set of practices. Using the construction of the English regions (North East, West Midlands, etc.) as an example, Painter demonstrates how a set of techniques and processes, such as the compiling of regional statistics, the creation of development agencies and the devising of regional strategies, confers a sense of delimitation, contiguity and coherence to the idea of regional divisions. Yet these apparently solid territorial blocs are created and sustained through complex networks of agencies and processes that transcend the boundaries of those regions.

Box 2.2 Stuart Elden on territory

While Sack drew attention to territoriality, Stuart Elden's recent work is an attempt to focus on the relatively under-theorized concept of territory. He argues that within debates on territoriality, the idea of territory itself has been taken for granted so that 'strategies and processes toward territory . . . conceptually presuppose the object that they practically produce' (2010: 803). Elden suggests that more attention needs to be placed on territory which is a somewhat more complicated concept than we might think. He argues that our contemporary ideas of territory emerged as a consequence of developments in cartography and spatial calculation. Consequently, he argues that territory reflects a way of thinking or conceiving of geographic space. Territory can be seen as something calculable, mappable and controllable. Furthermore, Elden also argues that, despite ideas of globalization and an attendant sense of deterritorialization (implying that borders are becoming less significant), territory remains a key component in contemporary global politics. Far from disappearing, he suggests that territory is continually being reconfigured and that spatial political relations are continually being shaped and re-shaped in ways which raise serious questions about the relationships between states, territory and sovereignty. More specifically, Elden (2007b) suggests that ongoing geopolitical events in Iraq, Afghanistan and elsewhere serve to demonstrate that territorial sovereignty is highly contingent on the behaviour of a state, but external political and military intervention against 'rogue' states is enacted alongside an apparent refusal to countenance any reconfiguration of state borders.

The creation of states and the contingent drawing of boundaries between neighbouring states represents the most obvious political expression of territoriality. Power is being exercised over that particular bounded space through systems of rules which govern entrance and behaviour. In this way, territoriality can be seen as the spatial form of power using a bounded space in order to secure a particular outcome (Taylor 1994), though we should be mindful of the fact that territories are not always rigidly bounded, or at least not in a formalized and conventional manner (Crampton 2011). While some of the most obvious (and contested) expressions of territoriality are manifested at the level of the state, many more micro-level examples of territorial control and territorial strategies may be observed. Territory and territoriality run through a wide range of social issues including race, class, gender and sexuality and this is reflected in the creation and sustaining of a vast array of spatial enclaves (Sidaway 2007). A street gang claiming space is a territorial reflection of the assertion of power. The demarcation of particular rooms within a house or workplace is also a form of territorial behaviour through which power (the power to include or the power to exclude) is expressed. For example, the prohibition of employees from certain rooms or areas within their place of work is an assertion of managerial power. In a similar vein, parental power is often expressed territorially through, for example, discouraging children entering some rooms in the house such as the kitchen or the parents' bedroom. Our need for personal space can also be seen as a form of territorial behaviour, the desire to have our own portable micro-territory (Hall 1959). Although these may be less obvious and may often seem more vaguely defined with less clear-cut boundaries, these 'informal' territories can still convey quite clear meanings to those concerned (Delaney 2005). Another key point here (and following logically from the previous comments) is that the bounding of space brings with it exclusion and control but human territories and their boundaries may be subject to periodic or continuous contestation, modification, transformation and destruction (Paasi 2008).

It is also worth pointing out that territory does not need to be intimately known in order for claims to be expressed. Laying claim to uncharted lands was commonplace during the era of imperialism leading to often unpredictable long-term consequences. A classic example of this was the Treaty of Tordesillas in 1494 which divided the world into Spanish and Portuguese territories, utilizing a line of longitude. Anything west of this line became Spanish, anything to the east could be claimed by Portugal (Figure 2.1). Essentially this was designed to give the Spanish free rein in the Americas and the Portuguese control of Africa. The subsequent discovery that part of South America lay east of the line accounts for the Portuguese colonization of Brazil. One legacy of this is that the predominant language in present-day Brazil is Portuguese; elsewhere, in central and South America, it is Spanish. This is also a prime example of major powers carving up geographic space between them with no regard for those who occupy those spaces. This example also serves a useful reminder of the importance of mapping territory. Maps are sometimes seen as neutral depictions of geographical realities but a more critical analysis suggests that maps have always been useful weapons in larger political projects such as the claiming of territory and in the maintaining of control over it (Harley 1988; Wood 1992; Black 1998). Mapping of territory itself functions so as to enhance power

Figure 2.1 Spanish and Portuguese hemispheres recognized by the Treaty of Tordesillas

sending out messages signifying control over portions of geographic space. Maps of the British empire conventionally depicted Britain's overseas territorial possessions in pink, conveying information but also proclaiming power over roughly one quarter of the planet's land area (Figure 2.2). Advances in cartography and the means of calculating terrain altered the ways in which space was considered. In turn, this led to the idea of territory and, ultimately, to attempts to apportion and control it. The military and political underpinnings of cartographic developments should not be ignored and the consequent role of mapping (in both practical and symbolic terms) in the creation of colonial territory was a key element in the imposition and maintenance of political control (Smyth 2006; Hewitt 2010).

Territory, identity and place

Even if territorial behaviour is seen largely as a phenomenon conditioned by our circumstances, it is readily apparent that people display a tendency to identify with particular places. Precisely what we mean by place remains extremely elusive but it is nevertheless apparent that people form bonds with place and these attachments may serve as an integral component of self-identity. Within a humanistic tradition, the importance of

Figure 2.2 British Empire

place and people's connection to, even love of, place has been emphasized (Tuan 1974). The strength of this tendency is clearly quite variable and may be more overt for some people in certain places rather more than for others. This 'sense of place' may be associated with places we like – the place we grew up in, places we have spent some time in, places with fond memories, places associated with positive experiences. Equally of course, people may often have rather different feelings towards certain places. Places in which bad experiences took place or in which tragic events occurred may be seen in quite negative terms. Whether place associations are positive or negative, what seems certain is that places 'are the focus of personal sentiments, with the feelings for place permeating day-to-day life and experience' (Muir 1999: 274). While modern communications technologies and associated developments may appear to suggest that geographic location is less important and that space has been annihilated by time, the reality seems to be that place most definitely continues to matter (Holloway and Hubbard 2001; Cresswell 2004).

The intrinsic geographic importance of territory is stressed by Soja, who suggests it 'provides an essential link between society and the space it occupies primarily through its impact on human interaction and the development of group spatial identities' (1971: 33). Once created, territories can become the spatial containers in which people are socialized through various social practices and discourses. As Paasi (2008) suggests, a number of important dimensions of social life and social power are brought together in territory. There is a material component such as land, there is a functional element associated with control of, or attempts to control, space and there is also a symbolic component associated with people's social identity. The social, cultural and political are brought together so that people identify with territories in such a way that they can be seen 'to satisfy both the material requirements of life and the emotional requirements of belonging' (Penrose 2002: 282). Although Sack suggested that territoriality 'forms the backcloth to human spatial relations and conceptions of space' (1986: 26), territory is more than a mere backdrop or the material manifestation of a set of social relations, it is also itself a form and effect of particular social relations bound into the intersections of power, space and society (Brighenti 2010a). The spatial is not simply the outcome of the social but the two are intrinsically bound up together (Delaney 2009).

While acknowledging the significance of place and its connections with identity, it is important not to overly romanticize these. Harvey (1996) warns of the dangers of the fetishization of the links between people and place whereby these come to be seen as 'natural' rather than conditioned via a variety of social, economic and cultural processes.

While people undoubtedly establish links (quite often deep-seated) with places, these are not necessarily the result of nature, nor are they explicable through reference to the mechanistic interpretations of some branches of social psychology (Dixon and Durrheim 2000). Rather, they reflect particular circumstances. Our love of certain places is as likely a function of events which have happened in that place as it is of any intrinsic values of the place itself. Furthermore, the 'naturalizing' of links between people and place, and the idealization of 'home' and 'community' can both reflect and engender a retreat into a conservative and reactionary parochialism. We might juxtapose constructs of place, territory and identity which are conservative, static and exclusionary with those which might be seen as progressive, dynamic and inclusive (Massey 1994). Tension centres on the promotion or retention of values of mutuality and solidarity while not retreating into a defensive insularity.

From the foregoing discussion it is clear that humans create, shape and re-shape territory. However, this is not simply a one-way process. Just as human behaviour results in the creation of territories so these territories, and the processes by which they are produced, impact on human life in a myriad of ways. People shape territories and territories shape people. In its most obvious form, economic, social and cultural practices vary across formalized territorial boundaries. In this way, such things as agricultural activities and settlement patterns may differ considerably between areas divided by a territorial boundary as a consequence of different laws. Cultural practices may differ on either side of long-established borders. Territories and associated territorial strategies can be seen as powerful influences which shape social life. Territories are more than simply bounded spatial entities, they can be seen as a fusion of meaning, power and space (Delaney 2009).

Summary

This chapter has presented a brief overview of how territories are produced and different ways of viewing the phenomenon of human territoriality. Territories emerged out of particular ways of conceiving geographic space. The desire to exert territorial control needs to be viewed in its broader social and political context, not be seen merely as a response to a biological urge. Territoriality, when viewed as a mechanism of power, is more than a simple behavioural trait. Territorial formation, control, resistance and transgression are all political phenomena. Territory and territoriality are important in shaping both our global and our local worlds. While it might be supposed that places and territorial boundaries are becoming less and less significant in an increasingly globalized world, the chapters which follow suggest that the territorial division of space seems likely to continue to be important for the foreseeable future. Place continues to matter, or be made to matter, and the (re)production of territories and the creation of boundaries remains an important phenomenon with many varied economic, social, cultural and environmental, as well as political, consequences.

Further reading

Many of the works highlighted in the chapter are listed below including some of the classic contributions on territoriality in the geographical literature. The list also includes some of the populist anthropological literature on territoriality and some more nuanced work. Important contributions on a sense of place and the connections between people and place are also highlighted.

Alland A. Jr., (1972) *The Human Imperative*, New York: Columbia University Press.

Ardrey, R. (1967) *The Territorial Imperative. A Personal Inquiry into the Animal Origins of Property and Nations*, London: Collins.

Cresswell, T. (2004) *Place. A Short Introduction*, Oxford: Blackwell.

Dawkins, R. (1976) *The Selfish Gene*, Oxford: Oxford University Press.

Elden, S. (2007) 'Governmentality, calculation, territory', *Environment and Planning D: Society and Space* 25 (3): 562–580,

Elden, S. (2009) *Terror and Territory. The Spatial Extent of Sovereignty*, Minneapolis: University

Elden, S. (2010) 'Land, terrain, territory', *Progress in Human Geography* 34 (6): 799–817.

Gottman, J. (1973) *The Significance of Territory*, Charlottesville: University Press of Virginia.

Gould, S. J. (1983) *The Mismeasure of Man*, London: Penguin.

Gould, S. J. (1991) *The Flamingo's Smile*, London: Penguin.

Hinde, R. A. (1987) *Individuals, Relationships and Culture: Links Between Ethology and the Social Sciences*, Cambridge: Cambridge University Press.

Holloway, L. and Hubbard, P. (2001) *People and Place. The Extraordinary Geographies of Everyday Life*, Harlow: Prentice Hall.

Ingold, T. (1986) *The Appropriation of Nature. Essays on Human Ecology and Social Relations*, Manchester: Manchester University Press.

Lorenz, K. (1966) *On Aggression*, New York: Harcourt, Brace and World.

Morris, D. (1973) *Manwatching. A Field Guide to Human Behaviour*, London: Jonathan Cape.

Morris, D. (1994) *The Naked Ape. A Zoologist's Study of the Human Animal*, London: Vintage, 1994. of Minnesota Press.

Painter, J. (2010) 'Rethinking territory', *Antipode* 42 (5): 1098–1118.

Rose, S., Lewontin, R. C. and Kamin, L. J. (1990) *Not in our Genes: Biology, Ideology and Human Nature*, Harmondsworth: Penguin.

Sack, R. (1986) *Human Territoriality: Its Theory and* History, Cambridge: Cambridge University Press.

Soja, E. (1971) *The Political Organization of Space*, Washington: Association of American Geographers.

Taylor, P. (1994) 'The state as container: territoriality in the modern world-system', *Progress in Human Geography* 18 (2): 151–162.

Tuan, Y.-F. (1974) *Topophilia: A Study of Environmental Perception, Attitudes and Values*, Englewood Cliffs: Prentice Hall.

3 The territorial state

As suggested in previous chapters, the division of the world into bounded political units, commonly referred to as states, is the best-known example of formalized territories. Quite often the terms 'nation' and 'state' are used interchangeably or in tandem as in 'nation-state'. However, this tendency serves to confuse rather than to clarify. Essentially the countries which form the world political map are states. These are legal and political organizations with power over their citizens, those people living within their boundaries. Although the two are clearly related, a nation is a somewhat more nebulous concept. It is a collection of people bound together by some sense of solidarity, common culture and shared history. Usually, this sense of a common identity is underpinned by a historic attachment to a particular territory or national homeland. In some instances, there may be a close approximation between nation and state. In France or Japan for example, the vast majority (though not all) of the state's inhabitants see themselves as French or Japanese. The term 'nation-state' serves to provide an impression of national and cultural homogeneity within the borders of a given state. However, all states contain within their borders a variety of nationalities. While the nation refers to a social collectivity bound together by elements of shared culture, the state refers to a set of political institutions which have jurisdiction over a specified territory. The state might be viewed primarily as a political entity with the nation as a more cultural construct. The concept of the nation and its importance in contributing to state formation and territoriality, together with the associated ideology of nationalism, are considered in Chapters 4 and 5. This chapter focuses on the state and its territorial dimensions.

The state is currently the world's dominant form of political organization. When we look at a political map of the world, what we see is a division of territory between these political units (Figure 3.1). It is tempting to view this as 'natural' and people's everyday thinking is

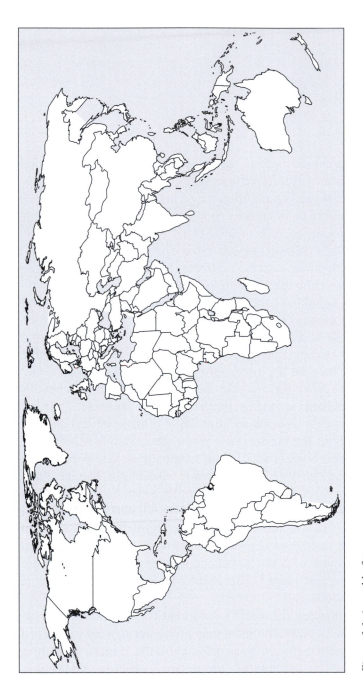

Figure 3.1 A world of states

heavily permeated by the 'taken for granted' presence of these states. Consequently, we tend to have a very state-centred view of the world. However, the state system displays considerable dynamism and in recent years the number of states has risen dramatically. In 1930, there were only about 70; now there are approximately 200. The collapse of communism in the early 1990s and the resultant break-up of the Soviet Union, Yugoslavia and Czechoslovakia led to the creation (or, in some instances, re-creation) of a number of new states such as Kazakhstan, Ukraine and Slovenia. In a very short space of time, a total of 22 states had replaced the previous three, with subsequent additions to this. East Timor (Timor-Leste) was recognized as a sovereign independent state in 2002 after decades of Indonesian occupation while both Montenegro and Kosovo have seceded from Serbia (in 2006 and 2008 respectively). In 2011, following a referendum, South Sudan became the world's newest state (seceding from Sudan) and Juba became the newest capital city. This chapter examines the state as a territorial entity and as a political unit. It explores theories of the origins of states before discussing the key features of states. Theoretical overviews of the role and functions of the state in the contemporary world are subsequently examined. The chapter concludes by exploring the issue of citizenship, the relationship between the state and those living within its borders.

Before proceeding, however, it is important to clarify another area of confusion – that between state and government. The state can be regarded as an enduring apparatus of power while governments can be seen as the agents which carry out the day-to-day running of the state. Elections in the United Kingdom in 2010 saw the Labour Party lose power to be replaced in government by a coalition of two parties: Conservatives and Liberal Democrats. While this represents a change of government, it leaves the state unaltered. The argument might be made that in totalitarian regimes (such as in North Korea) there is an overlap between government and state where the ruling party essentially constitutes the state. However, even in Eastern Europe, although some states fell apart, most of them (such as Bulgaria, Hungary and Poland) continued to exist after the demise of their one-party communist governments in the early 1990s.

Origins of states

As suggested earlier, states are not 'natural' entities, rather they are human creations. They represent a formalized division of the world into political units. Despite being the dominant political–territorial form, states have not always existed. Prior to coming into existence in their current form, early examples of states can be found throughout history. The best known of these include Ancient Egypt, Greece and Rome. Greece consisted of a series of city-states while Rome was the centre of a spatially extensive empire. In other parts of the world, state equivalents also existed. Examples include the empires of the Toltec, Aztecs, Mayan and Incas. These various entities waxed and waned but, while there might be some similarities with modern states, they were not based on ideas of sovereign territory as we currently understand it (Elden 2005). The concern here is with the origins of the modern state system which emerged as a sort of intermediate form between small city-states and larger empires. This is usually seen as an outcome of the Treaty of Westphalia in 1648 following the end of Europe's Thirty Years War, which helped set the scene for a territorial apportionment of land between European rulers. The subsequent territorializations can be seen as a politicization of space; space which could be measured and mapped, and which could consequently be controlled, apportioned and bordered (Elden 2007a). Westphalia led towards subsequent agreements over what land belonged to whom and, hence, it gave rise to the emergence of demarcated borders separating neighbouring polities. In considering how states have come into existence, geographers and others have been instrumental in developing theories of state formation and evolution.

Organic development of the state

Some geographers have propounded theories which view the process of state formation as one of territorial accretion. The German geographer Friedrich Ratzel likened the state to a living organism. He argued that states would expand territorially outwards in order to increase their size and power (Box 3.1). In so doing, they would devour smaller surrounding states in a struggle for space (Muir 1997). This serves as an excellent example of nineteenth-century geographers using biologically based ideas of territoriality, or the application of scientific thinking, to political–territorial issues (referred to in Chapter 2). In this way, a form of social Darwinism is used to explain political processes in territorial terms.

Box 3.1 Friedrich Ratzel

Ratzel was a German geographer who wrote an important series of geography texts in the nineteenth century. His ideas on the territorial expansion of states are linked to a school of thought known as environmental determinism. Essentially, this proposed that human behaviour was largely determined by the natural environment whereby climate and topography were seen as major influences on the way in which humans behaved. It is now widely accepted that people's behaviour, while it may well be affected by elements of the natural environment, is certainly not determined by them. Humans are subject to stimuli within the social environment and much of our behaviour is socially conditioned rather than biologically or naturally determined. Environmental determinism tended to bolster forms of scientific racism whereby people from certain places were judged to be inferior (Peet 1998).

Ratzel was also an ardent supporter of the German state and his intellectual ideas merged with his political views. He argued that states behaved as organisms and as such they needed to acquire more territory in order to remain strong and survive. He coined the term *lebensraum* to denote the idea of extra living space required by the state. Although he was dead before Hitler came to power, the idea of *lebensraum* was elaborated on by others and it eventually became central to Nazi ideology, helping to engender support for German territorial expansion in the 1930s, ultimately resulting in World War II.

Ratzel appears to have based his theory of state expansion and territorial control on Herbert Spencer's idea that human societies are social organisms. For Ratzel, just as every organism needed territory, human societies (as social organisms) required territory. Population growth led to pressures which required the acquisition of more territory or extra living space – what Ratzel termed *lebensraum*. Ratzel's ideas accord with those of other nineteenth-century thinkers, such as Friedrich List and John Stuart Mill, who argued that the ability of a territorially based community to attain or maintain nation status was dependent on its size in population terms (Hobsbawm 1990). It followed that a large population and an extensive geographical coverage were necessary prerequisites for statehood. However, given the increasing fragmentation along national lines in recent decades, and the longevity of tiny political entities such as Luxembourg, theories of the stability of states need to take into consideration other factors besides size.

From a functional perspective, these ideas espousing the perceived 'natural' or organic need for more territory served as a useful rationale for European colonial expansion. The need for territory was a convenient intellectual justification for overseas conquest. The extent to which geography is implicated in Europe's colonial past demonstrates that geographical and social Darwinist ideas were useful in providing a rationale for Britain and other European powers to engage in the acquisition of territories which have now become the present-day states of Africa, Asia and the Americas. Geography and geographers such as Halford Mackinder were actively implicated in this imperial project (Livingstone 1992; Ó Tuathail 1996). Mackinder is most remembered for his heartland theory, or geographical pivot of history, in which he suggested that, while sea power had served Britain well, technological changes, most notably the expansion of railways, meant that control over major land masses was of crucial importance. He argued the Eurasian landmass (centred on Russia) was of crucial significance and containing Russian expansion was of the utmost importance for Britain in its wish to maintain political prominence. Mackinder, like Ratzel, wrote from what can be seen as an imperial point of view. Another German geographer, Karl Haushofer, utilized and developed Ratzel's ideas of *lebensraum*. Hitler's subsequent use of the concept as a justification for German territorial expansion meant that both Haushofer and his ideas fell into disrepute. For all these 'imperial' political geographers, their ideas were bound up with the influence of both physical features, such as mountains and rivers, and human geographical considerations on politics and their role in affecting the strategies of states. Their task, as they saw it, was the devising of practical geo-strategies which could be utilized by political leaders.

State stability and coherence

State formation and stability can also be viewed in functionalist terms. Richard Hartshorne (1950) argued that states are under two types of pressure. The first of these are centrifugal forces. These are elements which act so as to pull apart or destabilize the state. These factors might include a larger land area. It is assumed that the larger the territory covered, the more difficult it is to retain control of it. Clearly, however, whether a state maintains its territorial integrity cannot be a function merely of size. It will depend on human features within its borders. Elements, such as the extent of regional inequalities, ethnic diversity, language differences, will be influential. In a state with sizeable regional differences, the poorer area may feel it is being exploited by the richer and may, therefore, wish to secede on the basis that it would be better off 'going it alone'. Alternatively, the richer region may view the poorer region as a drain on its resources. The end result, again, might be secession.

The ethnic diversity of the population might be regarded as a key element in determining the stability of the state with ethnically diverse states more likely to be unstable. However, this is only the case if a particular ethnic group feels it has been discriminated against and has been 'politicized' to the extent that it wishes to attain nationhood and thereby acquire its own political apparatus, namely statehood. Linguistic or religious differences (often associated with ethnic differences) can lead to pressures for separation. In Belgium, language is the principal political fault line with a division between Flemish speakers and French speakers (Walloons) which continuously threatens state coherence. However, the example of Switzerland as a stable multilingual democracy (there are four official languages) suggests that language differences do not always create political instability. The key element here is whether particular groups see themselves as constituting a cohesive body such that they feel themselves to be a 'nation' entitled to their own state. In general terms, it is argued, the closer the coincidence between nation and state, the more stable the state.

The second type of pressure to which states may be subject are centripetal forces. These serve to bind the state together. If centrifugal forces act as sources of instability, then centripetal forces have the opposite effect. They might include elements such as a common language, religious homogeneity, lack of extreme regional inequalities, and so on. Under such circumstances, it might be argued that there will be few pressures on the state as currently constituted.

Inter-state system

While ideas of state evolution and internal coherence have some merit, they imply that states develop as part of some inevitable process but this is not necessarily the case. There is nothing inevitable about the existence of France, or Italy or Spain. Many alternative states might have emerged, as some have in the past and others may well do in the future. The existence of the present-day state system is taken for granted and we see a political map which reflects what actually happened rather than what might have happened. Each state's initial existence is not subject to question. It is merely its size and stability which are up for consideration. The question still remains as to why some political configurations arose rather than others. We need to be wary of reading history (and geography) backwards.

In an inter-state system, the growth of states is at the expense of others which are either territoriality reduced (as with Serbia in recent years) or swallowed up entirely. While regional and ethnic tensions can have a significant bearing on the stability of a state, rather than a narrow focus on internal characteristics, relative power balances between states need to be taken into account when attempting to understand the evolution of a world-state system. Flint and Taylor (2007) suggest that Wallerstein's world systems theory allows state formation to be viewed from the 'outside' through the lens of inter-state relations rather than from the 'inside' with its reliance on states' internal characteristics. Such a perspective also allows us to see states as elements within a broader capitalist system, rather than viewing them in isolation. In addition, while conventional views of the state tend to regard it as a set of institutions freely entered into, this is usually very far from the case and states are more usually 'imposed' on people; as Lefebvre (1991) contends, states are born of violence. Much modern state-building in Africa originated with an external imperial power creating a territorially bounded unit to serve its own interests (Warner 1998). Rather than a seamless and unproblematic evolution, a state system has evolved in which issues such as contested sovereignty are of vital importance. In addition, competing nationalisms (which are not necessarily issues internal to the borders of existing states) and their claims for self-determination need to be looked at more fully and these are discussed in detail in the next chapter. In summary, the creation, maintenance and vitality of states has much to do with the balance of power between countries, rather than being simply reducible to internal characteristics.

Key features of states

While we may have a common-sense notion of what a state is, we need to delve a little more deeply to uncover its essential characteristics. The state is not a single unambiguous entity, rather we might think of it as a constellation of various elements and component parts. While precisely what it is may be hard to pin down, the modern state is dependent on two essential features: territory and sovereignty. Together, these two lead to the establishment of a third key feature – borders.

Territory

While the state is a political organization, it must have a territory – a portion of geographic space (land, air, water) over which it exercises power. States are endowed with the power to legislate for their territory and for those people living in that territory. Max Weber (1978) conceptualized the state as exercising a monopoly of power over the people living within its recognized borders so that 'only the state is inherently centralized over a delimited territory over which it has authoritative power' (Mann 1984: 69). Clearly, that power is sometimes contested, both internally and by external powers. Thus, the state can be seen in relation to four key elements – monopoly, territory, legitimacy and force. This means it is different from transnational corporations and other institutions of power which are not spatially restricted in the same way and which do not have the same array of powers (though they may still possess enormous influence). However, while we take the idea of territory for granted, it is, as suggested earlier, a comparatively recent concept. As Stuart Elden (2010) argues, territory is more than simply land (or geographic space), although that is important; our ideas of territory are dependent on ways of measuring and mapping it. Without developments in both cartography and geometrical means of areal calculation, we would not think of territory in the ways we do. With these advances came also legal developments associated with control of resources through control of territory. States emerged alongside the development of bureaucratic systems necessary for administration. Information gathering, both on the nature of the land, its features and resources, and on the inhabitants, through such means as censuses and other surveys, became essential components in state-building. Developments in cartography and information gathering led to a way of conceiving geographic space: to an awareness of what was 'ours' so that land was converted into territory – space which was owned, controlled or administered. These activities make the state legible for those who wield power within it and affect the way in which citizens perceive the state (Scott 1998). For imperial powers such as Britain, mapping their overseas possessions was about identifying resources and routes of access. It was also about making the territory visible, accessible, governable and placing a cultural stamp upon it. In Ireland, for example, the territory was claimed by Britain (though swathes of it were never entirely controlled), rendering it knowable through both mapping it and naming it in an anglicized way (Smyth 2006). More broadly, mapping has generally been bound up with military, political and economic imperatives – with taming and civilizing the savage 'other'. In North America, exploration and mapping of parts of the continental interior (most notably tracing routes through the Rocky Mountains) was bound up with trading and commercial interests (Nisbet 2007).

Territory is obviously important as land is a clear necessity for a state to exist but it also provides the state with a seat of power and a functional space in which the state operates (though its actions may extend well beyond its own territory). In addition, the state serves to inculcate a sense of territorial or national identity among the populace. This scenario is well summarized by Escolar:

> Gradually, all the emerging absolute states required cartographic techniques in order to represent their own geography, both for the purpose of the social legitimization of their dominions and as a means of broadcasting the geographical profile of the country throughout the realm and in order to draw up an inventory of their natural, social and human resources and to set up jurisdictional institutions to facilitate government and state administration.
>
> (2003: 32)

At an elementary level, we can say that states are territorial containers (though increasingly leaky ones) which have jurisdiction (or at least claim it) over politics, economy, society and culture within their acknowledged geographic spaces (Taylor 1994). However, we need to bear in mind that territory itself results from the processes of state evolution and the mechanics of mapping and calculating space. In a sense, then, territory only exists because the state does; territory is thus more than a feature of the state, it is also an effect of the system of power relations within which states exist (Brighenti 2010a).

Sovereignty

In addition to territory, there is one vitally important additional feature which states must possess: sovereignty. This is a much-debated concept but at its heart it refers to the authority of a state to rule over its territory and the people within its borders; that is, the right of the state to rule without external interference. Sovereignty implies the existence of geographic space over which control is exerted; territory is produced and claimed. However, declarations of territorial sovereignty in themselves are not sufficient. A state must be recognized by other states. In an inter-state system, if a particular state's sovereignty is not recognized, then its legitimacy is called into question. Sovereignty is therefore not just about self-declaration, it is contingent on its acceptance by external parties.

During the apartheid era in South Africa (when a racial ideology was used to consign black people to separate spaces), a number of *Bantustans* were created by the white minority government. These were ostensibly

The territorial state   41

devised as self-governing 'homelands' for black residents in what was essentially a charade designed to imply the existence of political autonomy for sections of the black population. These homelands, pronounced 'independent' by the South African state, were not sovereign because the international community refused to recognize them as such. Only South Africa viewed them (or claimed to view them) in this way. A number of contemporary examples of what can be described as quasi or pseudo states exist, many of them emerging following the collapse of communism in eastern Europe in the early 1990s. These include Abkhazia and South Ossetia, which have seceded from Georgia, and Transnistria, which has declared itself independent of Moldova (Figures 3.2, 3.3, Box 3.2). The Turkish Republic of Northern Cyprus was created following a Turkish invasion of Cyprus in 1974 as an apparent response to oppression of Turkish Cypriots by the majority Greek Cypriots. While these entities function as de facto states, they lack any widespread international recognition (Blakkisrud and Kolstø 2011; O'Loughlin *et al.* 2011).

Box 3.2 Transnistria

When Moldova, which lies to the east of Romania, became an independent country with the break-up of the Soviet Union, the narrow elongated piece of land on the eastern bank of the Dniester river declared itself independent from the newly created state. This self-declared Pridnestrovian Moldavian Republic (which also includes some pockets of territory on the west bank of the river) enjoys no international recognition, although it operates to all intents and purposes as a separate independent state (Figure 3.3). The Transnistrian 'state' claims a long cultural lineage that serves as a justification for the existence of an independent state. It sees itself as politically and culturally separate from Moldova, though the ethnic composition of its population consists mainly of Moldovans, Russians and Ukrainians. While it possesses the infrastructure and apparatus of a state, Transnistria lacks international recognition and only Georgia's break-away regions of South Ossetia and Abkhazia (themselves lacking effective international recognition) acknowledge Transnistria's independence. While Transnistria has endeavoured to project itself as open and democratic, it remains a highly authoritarian entity.

(See Blakkisrud and Kolstø 2011)

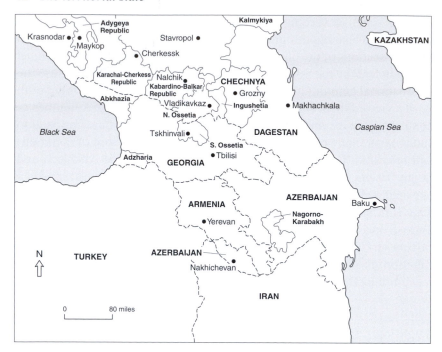

Figure 3.2 Caucasus region

Figure 3.3 Moldova and Transnistria

Even when such recognition is forthcoming, problems still emerge. The break-up of the former Yugoslavia and the creation of its successor states was sanctioned by the international community through the recognition of the sovereign claims of the newly created (or re-created) states. However, this recognition of sovereignty has of course been contested. Many Serbs were reluctant to concede territory to any of the non-Serb states, particularly Bosnia-Herzegovina and Croatia. Macedonia's independence also proved controversial due to Greece's concern that full recognition of an entity known by that name might lead to claims for an enlarged state incorporating part of northern Greece, which has a province called Macedonia. At present, the country is known officially (in the English language version of its name) as the Former Yugoslav Republic of Macedonia and is listed in atlases and gazetteers as FYROM. Kosovo's independence from Serbian rule is recognized by some 70 countries while others (not just Serbia) have not as yet recognized it as a sovereign entity. In addition, Kosovo's independence is very much dependent on the support of the United Nations and a range of other external bodies. For a nation to attain its own sovereign state it needs 'more than the strength of its national movement, the intensity of patriotic fervour or even its level of industrialization; it depends on politics and the constellation of power in the international system' (Harris 2009: 171–172). Questions may arise as to how many other countries must recognize the sovereignty of another for it to acquire 'recognized' status. Clearly, some countries may perpetually refuse to accord full recognition to the existence of others. Following the unification of Vietnam in 1975, after its war with the United States, it was many years until some countries recognized it as a sovereign entity. Indeed, the USA and UK did not do so until some 20 years later. The Federal Republic of Germany (West Germany) refused for many years to recognize the sovereignty of the German Democratic Republic (East Germany).

Sovereignty disputes often derive from disagreements over where borders lie. The African state of Eritrea attained independence in 1993 following a protracted 30-year struggle to secede from Ethiopia. Currently, the border between the two is disputed, with both countries claiming sovereignty over stretches of territory. In South America, Ecuador has a long-standing claim to sovereignty over part of northern Peru (Radcliffe 1998). A number of spatial entities have existed which might be said to exhibit, in part, the characteristics of states. The most obvious examples are colonies of imperial powers. While these might have had recognized boundaries, they were not independent sovereign entities remaining politically dependent on the external power. Some countries such as Britain still have a number of colonies, although the word 'colony' is generally no longer used to describe them. In the UK in the 1950s, the term 'Dependent Territories' was introduced; more recently this has been replaced by 'British Overseas Territories'. These are small entities, often islands such as Montserrat, Bermuda and the Cayman Islands, which have not as yet achieved full political independence from their colonizing power. Although Northern Ireland, Scotland and Wales have their own legislative assemblies, they remain part of the United Kingdom and are not fully sovereign entities.

Interestingly, despite their name, the individual states of the USA are not sovereign independent entities but, rather, are elements within a federated structure. Sovereignty resides with the United States of America. The same applied in the case of the former Soviet Union. Despite its title of Union of Soviet Socialist Republics, these republics were anything but sovereign (and far from socialist, many would argue). German *Lander* may have considerable autonomy, and their own regional governments, but they are not sovereign entities. The issues associated with enclaves and exclaves (areas belonging to a state but spatially separate from it, being embedded within the territory of another state) also bring with them questions of sovereignty. Although Nakhichivan is a landlocked entity lying between Armenia and Iran, it is neither independent nor a part of either of those states; instead, it is an exclave of Azerbaijan. While the region of Nagorno-Karabakh lies entirely embedded within the territory of Azerbaijan, with a predominantly Armenian population, it has been the centre of a long-running dispute, declaring itself a republic in 1991 (though it lacks international recognition) (Figure 3.2). The borderlands of India and Bangladesh contain many micro-territorial units: areas of land belonging to India but surrounded by Bangladesh and vice versa. These and other mini-units exist in some instances as remnants of feudalism, as a legacy of colonial break-up or the disintegration of larger territorial units. In any event, they raise interesting questions both for states and for their populations (Jones 2009). The everyday impact for people in particular places is further highlighted by struggles for control of territory in areas such as Colombia's Pacific coast region. Here a long-standing conflict between the government and left-leaning guerrillas has resulted in the territorial displacement of many thousands of people (Oslender 2007). Despite the contested nature of sovereignty in these and other instances, states are generally reluctant to cede territory even if they have little if any effective control on the ground.

In the current global geopolitical climate, territory and territorial strategies are deployed in intriguing ways, which raise interesting questions about the nature of sovereignty. In September 2010, it was reported that British military forces handed over responsibility for security in the Sangin area of Afghanistan to the United States. The fact that Afghanistan is neither British nor US territory (but has been under effective occupation since 2001) renders this a glaring example of the territorial exercise of political and military power. At a more micro-level, spaces of confinement and interrogation reflect complex territorial strategies. The US prison camp at Guantánamo Bay (on the island of Cuba) is *outside* the territorial United States and its ambiguous territorial status has proven an effective excuse in allowing the indefinite detention without trial of its inmates who are simultaneously subject to inter-rogation by US military authorities (Reid-Henry 2007). The extent of territorial control (rather than sovereignty per se) of US-led forces in Iraq and Afghanistan and the control of incarceration sites such as Abu Ghraib raise serious questions of legitimacy. This and other related practices renders the relationships between territory and sovereignty ever more complex and blurred (Gregory 2007; Elden 2009).

While debates rage over sovereignty with many contested claims, it should be borne in mind that it is something of an idealized condition rather than an immutable reality. Sovereignty is a political territorial idea that has evolved slowly and one which is almost always an ideal aspired to by states rather more than a fully achieved reality (Milliken and Krause 2003).

Boundaries

It follows from the discussion of territory and sovereignty that the emergent contemporary state system was associated with processes of boundary-making. If states have control over designated territory, it follows that there must be recognized boundaries separating their own territory from that belonging to neighbouring states. Clearly, boundaries may be contested; they may change over time or they can disappear, as happens when two states merge. The rise of the bounded state as a political unit necessitated a concern with the drawing and redrawing of political borders and the formalization of territorial arrangements. Events such as the Congresses of Vienna and Berlin in the nineteenth century represented attempts by political leaders of the world's major powers and their representatives to apportion territory to states and to (re)draw the borders between them in ways designed to reconcile larger strategic interests (Blacksell 2006). This imperial perspective is well encapsulated in the words of Lord Curzon, British viceroy in India at the start of the twentieth century:

Turkestan, Afghanistan, Transcaspia, Persia – to many these words breathe only a sense of utter remoteness, or a memory of strange vicissitudes and of moribund romance. To me I confess they are pieces on a chessboard upon which is being played out a game for the domination of the world.

(cited in Kleveman 2003: 3)

Subsequently Curzon, by then British Foreign Secretary, was involved in the post-World War I repartitioning of Europe at the conference of Versailles. At that same conference, US geographer Isaiah Bowman, as part of the US delegation, was similarly instrumental in the reconfiguring of Europe's internal borders (Smith 2003). While considerations of physical features (such as rivers and mountain ranges as 'natural' frontiers) and cultural characteristics of populations entered into such decisions, these major conferences can primarily be read as responses to the geo-strategic concerns of the larger powers and were predicated on military and political considerations.

Borders are places where the paraphernalia of the state is clearly manifested with flags flying at crossing points, signs welcoming the visitor to another country, passport checks and police controls. Here is made clear the distinction between 'our' territory and that of others. Traditionally, political geographers have distinguished between 'natural' boundaries and 'artificial' boundaries. The former category encompassed naturally occurring phenomena, such as rivers and mountain ranges. The Rio Grande river forms part of the border between the United States and Mexico, while the Mekong river forms part of the boundary between Laos and Thailand. Lines of latitude are seen as examples of artificial boundaries. Thus, the 38th parallel forms part of the border between North and South Korea, while the 49th parallel forms the border between the USA and Canada. However, the distinction between natural and artificial boundaries is somewhat misleading given that borders, whether utilizing naturally occurring features or not, are intrinsically artificial. States are human creations; it follows that their borders are, therefore, artificial. As a consequence of political decisions, some rivers become borders (or cease to be borders), others do not. Furthermore, these sorts of boundary classifications are also predicated on an assumption that borders must exist, the only debate being over where the lines should be drawn. It is assumed that such things as ethnic or national identity are fixed and unproblematic entities and that territorial divisions separating them are 'natural', desirable and attainable (Fall 2010).

The more traditional geographer's concern with a classification of boundary types is of much less significance than an exploration of the impact of the imposition of borders. Borders are constantly subject to change resulting from disputes over territory. These disagreements may relate to the precise location of the border, or they may centre on whether or not a particular border should exist in the first place. Numerous examples of changing borders exist. Bulgaria's borders shifted on a number of occasions between 1878 and 1947, when its present-day borders were agreed upon (Glassner 1993). The German–Polish border has been subject to debate and the present one was only finally accepted following German re-unification in 1991. The boundary between France and Germany has also altered during the course of this century as a consequence of both world wars, with part of Alsace-Lorraine switching between countries, while similar changes occurred along the Belgian–German border. The division of the Karelian region, bisected by the border between Finland and Russia, continues to be seen as problematic by some (Paasi 1999). Some borders are subject to ongoing dispute, such as those between Venezuela and Guyana, and Surinam's borders with both Guyana and French Guyana. This is a legacy of a division of colonized territory between European powers when the areas in question had not been fully 'explored' by Europeans and, hence, were never accurately mapped. In Europe, issues have arisen between newly created states. The border between Macedonia (declared independent in 1991) and Kosovo (declared independent in 2008 though not fully recognized internationally) has not been formally agreed and for much of the twentieth century was not an international border as both entities were part of Yugoslavia.

While the precise location of a border may give rise to conflict, in other instances the very existence of the border is what is in dispute. The political conflict in Ireland revolves around the existence of the border between the Republic and the North (Figure 3.4). For Irish republicans, who wish to see a unified Ireland, the border is at the heart of the conflict. It is seen as an imposed and artificial boundary whose removal is essential. In this situation, the border became a key site within the conflict. Until the late 1990s, it was a heavily fortified area which had witnessed many violent episodes throughout the conflict with attacks on British checkpoints and customs posts.

While some borders are relatively uncontroversial and others are hotly disputed, it is important to note that the relative significance of certain borders changes over time. The Italy–Yugoslavia border was once of huge political import, separating an EU member state from communist Yugoslavia. Now Italy's border with Slovenia (one of Yugoslavia's successor states) is comparatively insignificant with relatively free movement between two EU members. By way of contrast, Slovenia's

Figure 3.4 River separating the Republic of Ireland (on the left) and Northern
Ireland (on the right) (Source: Author)

border with Croatia (previously an internal divide between two component parts of the federal Yugoslav state) is now not just an international
boundary but also a EU frontier.

While borders are clearly political constructs, they have very real social
and cultural implications, particularly for those living in border zones.
Borders are not just lines dividing territory, they are social and discursive
constructs which have important ramifications – not just in a broad
political sense but also in people's everyday lives (Newman and Paasi
1998; Paasi 1999). Studies in various parts of the world, summarized by
Prescott (1987), suggest that borders often cut through areas which are
culturally homogenous, or at least in which there are no significant socio-
cultural differences either side of the line. However, the creation and
existence of formalized borders may lead to significant differences on
opposite sides (Rumley and Minghi 1991). For example, differences in
agricultural practices, levels of urbanization, or income levels, may
emerge as a consequence of differing government policies. Contrasting
landscapes may evolve in terms of phenomena such as different settlement patterns. People may also develop different attitudes towards the
border depending on whether they see it in a positive light and/or depending on whether they see themselves as being on the 'right' side of it. In

many respects, the creation or imposition of a physical border can result in the creation of a partitionist mentality through which people on opposite sides drift apart due to their political separation. Borders may serve to essentialize national identity and harden linguistic and cultural divisions. The evolution of a sense of difference between 'east' and 'west' Germans might be seen as a consequence of that country's political division between 1945 and 1991, a division marked by a heavily patrolled border exemplified by the construction of the Berlin Wall in 1961. Similarly, other national borders which served as Cold War boundaries, such as that between Austria and Hungary, had profound impacts on those living in close proximity to them (Meinhof 2002). Political decisions taken in remote political centres can have dramatic effects on people who find their lives disrupted by shifting borders. Groups such as the Lezghi in Dagestan (part of the Russian Federation) and Azerbaijan now find themselves divided by an international border as a consequence of the disintegration of the Soviet Union, leading to difficulties in maintaining a coherent identity (Birch 2007). The demise of the Soviet Union has also had serious repercussions for the Wakhi and Kirghiz peoples who find themselves separated by the borders of Kyrgyzstan, Tajikistan, Afghanistan, Pakistan and China (Kreutzman 2007). That borders have such an impact is, in part, the reason they are so often centres of conflict. Far from simply being lines on a map, they can have profound impacts on people's lives, as Oliver August's account of the division between East and West Germany graphically illustrates.

> My father had been fourteen when the war ended and the Allies drew a line across his father's tree nursery. The main house was in the Soviet zone while some of the fields were in the British zone. The border literally divided the property. Aged seventeen, my father hid a suitcase on a horse-drawn cart and drove west across the border on family property, leaving his parents behind. In the following forty years he was allowed to return only twice – for a maximum of three hours each time – for their funerals
>
> (2000: 3)

In a myriad of ways, borders become elements within people's everyday lives and shape their day-to-day being. The significance of borders stretches from the global importance attached to the Berlin Wall in the later part of the twentieth century, to the local but very real separation of people in cities such as Nogales and other towns and cities along the US–Mexico border (Figure 3.5). In addition, it is worth pointing out that for many the existence of state boundaries is, in some senses at least,

Figure 3.5 US–Mexico border wall (Source: Fernando Moleres)

largely meaningless. Many nomadic groups such as Bedouin people living in Iraq and Jordan tend to ignore (or at least try to) what to them are meaningless lines on a map. Similarly, gypsies in many European countries may regard state borders as of no great significance in terms of identity, having no strong allegiance to the state in which they live. This is not to suggest that borders are of no significance to these groups in other ways. Their treatment may be better in some states than in others, while the existence of political borders has consequences for their ability to move freely between states (Fonseca 1996). Rao (2007) suggests that peripatetic groups have acted as social and economic links between different places but such linkages are disrupted through restrictions on movement. For those living in borderlands, their lives are disrupted by these imposed divisions of political space and for people in contested zones, such as Kashmir or the Fergana Valley in central Asia, state efforts at defending territory (and national pride) have very real impacts (Megoran 2006; Box 3.3).

Box 3.3 Fergana Valley

The Fergana Valley is a large region in central Asia which incorporates areas straddling the borders of three former Soviet central Asian republics – Uzbekistan, Kyrgyzstan and Tajikistan (Figure 3.6). Having been incorporated initially into the Russian empire in the nineteenth century, the region became embedded in the Soviet Union in the twentieth century. With the demise of the Soviet Union, the region became divided among the successor states. However, state borders do not neatly coincide with ethnic divisions. The borders of the three countries entwine in an intricate manner and many people find themselves ethnic minorities within the state in which they reside. Since independence, sporadic outbreaks of inter-ethnic conflict have continued to occur. Each country also possesses exclaves surrounded by the borders of the neighbouring states. Ethnic unrest, coupled with periodic border closures, exacerbate tensions and make daily life difficult for many. Violence in the Kyrgyz city of Osh in 2010 led to many thousands of ethnic Uzbeks seeking refuge in nearby Uzbekistan. While the emergence of separate states and increasingly solidified boundaries has to some extent tended to reinforce ideas of separate nations, for those living in border areas their imposition may be rejected and contested (see Megoran 2006).

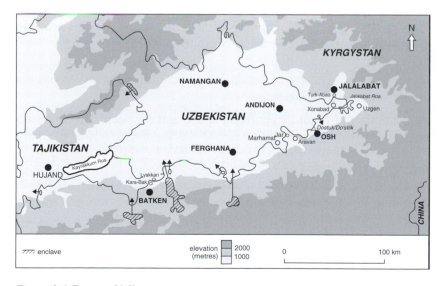

Figure 3.6 Fergana Valley

The role and functions of the state

Up to this point, aspects of state formation and conflict between states have been explored. This reflects the idea of the state as a geographical entity. However, the state is primarily a political unit and questions arise concerning what it is that states do. What is the function of the state? At its most basic level, it can be said that the state provides a legal framework, infrastructure and services to be used for the benefit of its citizens. More specifically, the state regulates the economy (although currently dominant economic theories suggest that states should minimize the exercise of this function). It also provides public goods such as health and transport services (although in many countries these services are increasingly privatized). It provides legal and other frameworks designed to guide and regulate its citizens' behaviour. The state also defends its territory and its people against external aggression and internal threats.

Viewed more broadly, it can be suggested that the state ensures (or may claim to ensure) representation, security and welfare for its citizens. People come into regular contact with the state. In most countries letters are delivered by state-employed postal workers, while medical, education and other services are often provided by the state. Streets are policed by a state police force. Many utilities are provided by the state. Even where service provision is by private companies, these are usually subject to some form of state regulation. In this way, there is a series of daily bonds between the state and its citizens. The state becomes a routinized element in people's day-to-day existence and it is reproduced through a variety of seemingly mundane practices. Flags being flown on government buildings or the reproduction of letterheads on official government stationery are among the 'ordinary' ways in which the state is reproduced and embedded in our consciousness. The state permeates our lives through various systems of regulation and control, evident in both our public and our private dealings.

In many countries, the period since the late 1970s has witnessed policies of de-nationalization whereby many state services have been privatized. This is part of a broader neo-liberal economic agenda whereby the state's role is seen as a regulator of services (at best) rather than a provider. The free market and private enterprise are assumed capable of providing the majority of public services. In the UK, services such as water supply and rail transport (previously supplied by state agencies) are now operated by private companies rather than by the state. This policy of 'rolling back the state' can be said to have had a definite effect in reducing the amount of direct contact people have with 'agents' of the state. Despite this, however, the state remains quite pervasive. Even privatized utilities operate within a state-imposed regulatory framework. In the so-called private domain, the state is virtually omnipresent, legislating for the age of sexual consent, the age at which alcohol can be consumed, registering births, marriages and deaths and embedded in other aspects of people's lives. The state remains strongly interventionist, maintaining a highly regulated framework through which civil society is territorialized (Mann 1984).

The extent to which the state is, or should be, involved in service provision and regulation is hotly contested. On the right of the political spectrum, it is argued that the state should interfere as little as possible with people's everyday lives. Intriguingly, this argument tends to relate primarily to welfare and service provision rather than to support for the private sector through favourable tax regimes, state loans and so forth. Within the United States there is a long and deeply entrenched view that the state should be a kept to a minimum. The reactions to attempts by the Obama regime to introduce a form of health care insurance, following his election as president in 2008, provoked intense debate with many arguing that such a scheme represented a completely unacceptable form of state intervention. On the left, it is generally argued that the state should play a key role in protecting its citizens and in regulating the territorial framework in which they live so as to ensure some degree of equality of treatment and protection for more vulnerable groups. The old socialist slogan of care 'from cradle to grave' epitomizes this perspective of the protective state. Dominant neo-liberal perspectives have, however, led to the 'hollowing out' of the state as private sector agents now carry out functions previously seen as the preserve of state agencies. The ongoing global financial crisis which has unfolded in recent years has somewhat dented belief in the supposed self-regulatory abilities of the private sector and has re-awakened concerns over rampant capitalism. It appears evident that a system of extremely light state control over the activities of major financial institutions has had far-reaching global repercussions.

Theories of the state

There are many different types of state. There are unitary states where power is highly centralized and there are also federal states with a high degree of political devolution to territorial sub-state units. Similarly, there are authoritarian states with quite rigid control and little popular participation, while there are more open state systems, symbolized for many by the liberal democracies characteristic of Western Europe. Historically, states have evolved from largely centralized entities with little popular participation through the development of electoral systems and gradual

extensions of the franchise, thereby extending formal political engagement from an elite downwards to include ever larger swathes of the
populace. While it is tempting to see this in evolutionary terms, it is
important to remember that most extensions of the democratic system
have resulted from social and political struggle resulting in achievements
such as voting rights for women, for example. At a simplistic level, there
is a tendency to think of the relationship between the state and its citizens
as a freely entered into set of arrangements. Clearly, in authoritarian states
this is most definitely not the case. However, even in so-called liberal
democracies, in which electoral systems are reasonably free and trans-
parent, there is a need to delve beneath the surface appearance of the state
as the repository of the collective will of its citizens in order to devise
adequate theories of the role and functions of the state. This section
summarizes the main types of theory relating to the state in liberal
democracies. In broad theoretical terms, there are two main ways of
conceiving of the state. These are pluralist theories and more critical
theories. More attention is devoted to the latter as its particular strength
is its usefulness as a critique of the pluralist perspective.

Pluralist perspectives

From a pluralist perspective, the state within liberal democracies is
viewed as a neutral arbiter. It is seen as being above, and separate from,
any vested interest. It is deemed to be apolitical in the sense of having
no interest in the form of society. Rather, it is seen as an institution which
is shaped according to the citizens' will, democratically expressed. This
reflects the basic democratic underpinnings of liberal democracy. People
vote in elections and sovereignty is seen to reside with the citizens. The
notion of popular sovereignty, whereby power emanates (via the electoral
process) from the people, derives from the philosophy of Hobbes and
Locke. People are deemed to be active participants in the process of
societal regulation. From this, it follows that governments merely act as
the agents of the electorate. People transfer their power to their elected
representatives who then act in their name, and order affairs of state
accordingly.

The classic liberal view of the state is that of the guarantor of the rights
of the individual. This rests on the notion of a clear private sphere which
should be protected from outside interference. The state, or public sphere,
is seen as necessary in order to safeguard this private sphere though,
ironically, debate then surrounds the apparent interference of the public
sphere within the private. In other words, the state is sometimes perceived
as 'a potential threat to the liberty it exists to secure' (Schwarzmantel

1994: 42). Liberal democratic states attempt to reconcile the classic liberal position of non-interference and freedom with the democratic ideal of popular sovereignty.

Pluralist states are seen as the polar opposites of totalitarian ones. Instead of one imposed view, there is a reconciliation of a variety of diverse views. Although at any one time a particular political party may hold power, and although they may represent a particular vested interest, the nature of the pluralist state is such as to ensure a system of checks and balances which will safeguard the citizenry against abuses of power. In this way, an independent judiciary, media, etc., are seen as integral components of the state apparatus. No one faction can have total control. States, on balance, are seen as neutral and as responsive to the various pressures which emerge within society. The state acts as a neutral entity mediating between different sections of society and the various elements of the broad democratic system are deemed to bring about appropriate solutions to the conflicts which arise. The state delivers the policies and services needed by society and is thus seen largely as a technical apparatus rather than an overtly political entity.

Critical perspectives

Although contemporary liberal–democratic states may appear to be built on notions of widespread public participation, some authors have pointed to the very 'political' nature of the state, indicating how, contrary to the pluralist view, debate is steered and shaped in particular ways. At a very basic level, it is suggested that many key issues never enter the arena of open political debate. The production and maintenance of nuclear weapons in Britain and the presence of US military air force bases in the country are two related examples of phenomena which have tended not to feature as election issues and are not subjects of parliamentary debate or public discussion This is a phenomenon Bachrach and Baratz (1962) have referred to as 'non-decision-making' through which, it is argued, many issues are kept off the official agenda and do not enter into the decision-making process. The UK actively engaged in invading and occupying Afghanistan and Iraq in the early 2000s but opinion polls suggest the majority of UK citizens oppose this. The endorsement of these acts of territorial occupation by the British government was not really a reflection of the democratic will of the people.

These suggestions of the steering of a certain agenda indicate that the state seems far from neutral; instead, it acts in the interests of specific sections of society. Rather than being an impartial referee 'the state constitutes an uneven playing field which privileges some forces and interests while proving less accessible to others' (Hay 1996: 7). The Italian theorist Gaetano Mosca suggested there is a distinction between a ruling class and a class that is ruled. The argument here is that there will always be a stratum in society who will tend to rule in the interests of the rulers rather than the ruled. Self-interest will take precedence over any broader concerns. This rests on a sociological distinction between an 'elite' on the one hand and a 'mass' on the other. In a variant on this, Joseph Schumpeter argued that people choose between competing elites. In this way, there is not, strictly speaking, a pluralist state but rather a constrained choice between various aspirants to power – what can be described as democratic elitism. Essentially, some people (those with a better education, access to the media, better connected into networks of power and so on) are in a better position to influence the manner in which the state operates. The sense in which members of parliament in most countries come from specific social backgrounds or that candidates for the US presidency tend to be rich (and can therefore mount expensive electoral campaigns) can be cited as evidence for this elite system of state control. Critics of elite theory argue that it is predicated on the idea of an innate tendency towards elite formation which leaves questions concerning the emergence of elites and whether their existence reflects a natural tendency.

Marxist theories of the state also suggest a division between rulers and ruled but one that is premised on the idea that societies are divided along class lines. There exists a capitalist class (those who control the means of production, in Marxist parlance) and a working class (or proletariat) who sell their labour to the capitalist class. The surplus value (profit, essentially) gained by the capitalist class as a consequence of employing the workers is seen as resulting from the exploitative nature of the relationship between these two classes. While most Marxist theoreticians would agree that the contemporary world is more complex than in Marx's day, nevertheless the basic foundations of the analysis remain the same. The distinction between a powerful class and a relatively powerless one is the basic fracture in capitalist society. They argue that what is important are the deep-seated structures within society: the hidden dimensions of power. Within this, class structure is seen as the key element and the state is viewed as a mechanism which acts in defence of that structure.

Far from being a neutral mediating party between various vested interests, the state acts in the interests of the prevailing capitalist system. Viewed this way, the state is not an independent structure standing apart from the rest of society. On the contrary, the state is deeply embedded within socio-economic relations (Held 1989). Marx himself did not devise a theory of the state. In *The Communist Manifesto* of 1848, Marx and Engels likened the state to 'a committee for organising the common affairs of the whole bourgeoisie' (1969: 44). In other words, in orthodox Marxist terms the state is seen as an agent of the ruling class, its existence necessary to deal with capitalist crises. It defends and protects the class structure by preserving existing power imbalances.

While the world may be somewhat different now compared to the era in which Marx and Engels lived, contemporary theorists following a Marxist tradition have endeavoured to theorize the state in terms of class divisions. It can be argued that contemporary states are still characterized by class conflict, even if the members of those classes are not always aware of it. The role of the state is to manage that conflict and to ensure the reproduction of the existing social order; in other words, to preserve the current dominance of the capitalist class, thereby ensuring the reproduction of the capitalist system and capitalist social relations; in short, to preserve the existing social order.

Liberal democracy tends to be accorded an elevated status as the pinnacle of political achievement. However, despite outward appearances, states are not necessarily the open democracies they purport to be. As Miliband argued in relation to the 'new' states of eastern Europe following the collapse of communism:

> the notion that the battle for democracy has already been won in capitalist-democratic systems, save for some electoral and constitutional reforms at the edges, simply by virtue of the achievement of universal suffrage, open political competition and regular elections is a profoundly limiting and debilitating notion which has served conservative forces extremely well, and which has to be exposed and countered.
>
> (1991: 13)

A debate between Ralph Miliband and Nicos Poulantzas in the 1970s demonstrates different ways of conceiving the state. Both viewpoints represent a strong critique of the pluralist perspective. Miliband adopted what is known as an instrumentalist view. From this line of reasoning, the state is argued to be capitalist by virtue of the class background of key figures. Taking the example of the United Kingdom, Miliband studied the origins and background of key power holders in society – not just politicians, but also members of the judiciary, media barons and other powerful figures. His argument is that the majority of power-brokers shared a middle-class background. In the main, he argues, this results in the state being ultimately controlled by a dominant elite. There is a concentration of power within a particular stratum: a coterie of people from relatively affluent backgrounds educated at Oxbridge colleges and so on. While subsequent governments have to some extent displayed a wider class composition, and there appears a greater spread in the social origins of opinion formers, it is still the case in the UK and elsewhere that the working class are underrepresented in positions of power, both formal and informal.

Poulantzas did not disagree with Miliband's assessment of the class-biased nature of the state elite. However, he felt that this sort of instru-mentalist analysis was not really necessary. Instead, he argued that the state is capitalist because it operates within a capitalist mode of pro-duction. It is this which is the determinant, not the class composition of the elite. The state is constrained: it must conform to the needs of capital. Even if members of the proletariat occupied positions of power, the state would remain capitalist because it is integrally embedded within a global capitalist system. The current global financial crisis provides ample evidence of the nature of the state as a tool of capitalist regulation. The banking collapse of 2008 has been followed by government bail outs in various countries in order to prevent the total collapse of the banking system. Not only have many leading bankers (whose actions precipitated the crisis) continued to reap huge rewards but, more significantly, the architecture of the global financial system has also remained largely in place. In the case of one of the worst affected economies, Ireland, an EU and IMF bail out was designed to shore up the Irish banking system (allowing Irish banks to repay their foreign creditors). In addition, the Irish government, while ushering in swingeing cuts to social welfare and a series of measures disproportionately impacting on the poorer sections of society, categorically refused to increase rates of corporation tax. This is a very clear defence of the edifice of the capitalist state. The state and capital are far from being separate entities; rather, David Harvey (2010) refers to the state–finance nexus as the 'central nervous system' of capitalist accumulation. Ultimately, the pluralist/non-pluralist distinction is one between the view of the state as a disinterested and disembodied regulator or administrator as opposed to an agent which reproduces society, culture and polity.

A final point worth noting here is that many states exhibit healthy signs of democracy when judged on a procedural basis reflected in such things as elections and judicial processes. This leads to an assumption that the state is highly democratic and has not been 'captured' by any vested interest. Indeed, the administrative apparatus of the state and its seemingly banal bureaucratic procedures tend to mask its more overt political role, thereby serving to reproduce it (Sharma and Gupta 2006). When judged in more substantive terms, the outcomes of the enactment of democracy may fall short of what the procedures suggest in terms of inequality and various forms of social exclusion. In short, procedural democracy does not necessarily translate into substantive democratization (Bell and Staeheli 2001).

The hegemony of the state

The role of the state, as outlined above, gives rise to an obvious question. If the state is 'biased' against the interests of the majority of its own citizens, why don't those citizens oppose its existence? Why does a revolution not occur? If it is the case that the state reflects certain dominant interests which are antithetical to the interests of the mass of citizens, then why does it persist? One way of explaining the resilience of the capitalist state, despite its apparent contradictions, is through the utilization of Antonio Gramsci's (1971) ideas of hegemony. Here, it is argued that the political, intellectual and moral leadership of a dominant class results in the dominated class actively consenting in its own domination. This is a result of the exercise of two different but related forms of control – the coercive apparatus and the ideological apparatus. In Mann's (1984) formulation, the distinction here is between despotic power (reliant on coercion) and infrastructural power (which is based on forms of negotiation). The first of these refers to institutions such as the police, army and judiciary. These serve as forces which keep people in line in a very obvious manner. The existence and power of this coercive apparatus is readily apparent in totalitarian states, yet it is no less important in liberal democracies, as evidenced periodically in the policing of protest marches and so on (Box 3.4).

Box 3.4 The spaces of political protest

Political protests such as marches, sit-ins or meetings take place in specific places. However, state authorities can delimit (or try to) where and when these may take place. Protest marches such as those against the wars in Iraq and Afghanistan have taken place periodically in many cities such as London. On these occasions, permission to march through the streets (public space) is granted though routes must be pre-agreed. Protests by students against increases in university tuition fees led to demonstrations in London and other British cities in 2010 and 2011.

Protest of a more permanent nature has generated controversy in Parliament Square in London. Its proximity to the UK parliament renders it a popular space in which to protest. Anti-war protestor Brian Haw set up camp there in 2001. He remained there until shortly before his death in 2011 and various other individuals and groups have also occupied space there for considerable periods of time (Figure 3.7). While some regard this as a spatial manifestation of a legitimate right to political protest, others tend to focus on the presumed illegality and the supposedly negative impact the 'residents' have on the appearance of the space.

Figure 3.7 Peace camp, Parliament Square, London (Source: Author)

However, of considerably more importance, especially in liberal democracies, is the much more subtle ideological apparatus. In Gramscian terms, there is a widely accepted set of 'common-sense' propositions that are rarely questioned. Gramsci referred to 'the "spontaneous" consent given by the great masses of the population to the general direction imposed on social life by the dominant fundamental group' (1971: 12). This is akin to what Foucault means when he says 'each society has its regime of truth, its "general politics" of truth: that is, the types of discourse which it accepts and makes function as true' (1980: 131). This acceptance of a particular way of viewing things is enabled through a variety of filters. Most obvious among these are education and the mass media. It is argued that these act as 'gatekeepers' of ideas, serving to permit discussion of some topics while largely ignoring others. In helping to shape public opinion, they can have a hugely conservative impact, protecting the state against potential opposition. The dominant ideas in society come to be seen as 'common sense' and alternative ideas are seen to be threatening, or unrealistic, or bizarre, or unworkable. For example, following the spectacular banking collapses of 2008, the repeated mantra, presented as self-evident and without need of justification, in the UK and many other economies is that spending cuts are necessary to ensure economic recovery. Alternative views that other courses of action might be pursued are dismissed. In addition, some areas are seen as immune from cutbacks, such as the building of more and more advanced weaponry to facilitate overseas military interventions. It is taken as obvious that money needs to be spent on certain arenas while others are seen as fair game for reductions in state support. The UK government engaged in military intervention in Libya in 2011 while simultaneously cutting spending in areas such as education. While newspaper headlines target welfare fraud, tax avoidance by the rich is considered almost 'natural' and inevitable. Particular policies or courses of action are seen as 'obvious', or the 'right' ones, or the 'only way' – strategies which all 'reasonable' people would support.

From a Gramscian perspective, this subtle ideological control is bolstered through systems of social welfare, provision of unemployment assistance, pensions, and other forms of support. Although these may be subject to periodic reductions, through these devices the state manages to deflect much opposition. They provide sufficient incentives to 'buy-off' the otherwise disaffected. The state is seen to provide certain services as a sop to placate the masses. In this way, welfare programmes and other attempts to provide some form of assistance to the less well off might be seen less as genuine attempts at relieving suffering and easing disadvantage, and more as efforts at maintaining political stability. The end result is that the political order is maintained so as not to undermine the basis of the state.

In a different way, various threats are invoked in order to maintain political control. In recent years, British and other governments have emphasized the somewhat generalized threat of terrorist attacks as a justification for both foreign and domestic policies: of overseas military adventurism on the one hand, and increasing domestic surveillance and curtailment of civil liberties on the other. While discourses of security abound and claims of terrorism and its prevention are invoked, it could be said that state hegemony is such that state forms of terror proliferate under cover of a range of fears (Pred 2007).

Within this discussion, a note of caution is necessary. It is important to bear in mind that states are not unified, self-contained, all-powerful units, but rather consist of a series of diverse and fragmented institutions whose impacts on society are hugely variable (Smith 2009). While the state obviously wields considerable power, spaces continue to be created in which opposition can be voiced.

Citizenship

Finally, in this chapter it is useful to discuss the relationship between the state, as a territorially based apparatus of power, and the people over whom it exercises that power – its citizens. Citizenship concerns the duties, obligations and rights of individuals in terms of their membership of a specific polity. People have a relationship with the state whereby they have duties to the state in return for which they enjoy certain rights. Citizenship is conferred by the state and it is not a static concept. States can, and do, alter their citizenship criteria. Although the concept of citizenship has evolved over time, many of our ideas on the nature of it stem from the philosophers Thomas Hobbes and Jean Jacques Rousseau. The former saw the relationship between people and the state as one where individuals subordinated themselves to a state and allowed it (or its government) to act on their behalf. Rousseau (1973) enunciated the idea of the 'social contract' whereby people gave their consent to be governed in return for the protection of the monarch or an equivalent form of government. There is obviously a trade-off here. Citizens will be protected by the state in return for giving their allegiance to it. It is this balance which permits societies to reduce a person's rights if that person is found to have failed in their obligations. Thus, people who commit crimes are seen to transgress that boundary through breaching the laws of society and the sanctions that are imposed may, depending on the nature of the crime, result in a reduction of their citizenship rights. For example, if a person is sent to jail, his or her freedom is obviously curtailed. He or she no longer has freedom of movement (Faulks 2000).

One of the clearest statements of the modern concept of citizenship is that made by T.H. Marshall (1950) who distinguished between three different, though related, sets of rights. These were civil rights (such as freedom of speech and travel), social rights (such as a basic standard of living and adequate health care), and political rights (such as the right to vote and to engage in political activity). Many countries outline formally the rights their citizens can expect. Possibly the best-known statement of international rights is the *Universal Declaration of Human Rights* adopted by the United Nations in 1948. Through the *Declaration,* the UN endeavoured to produce a common set of rights to which all countries should adhere. It states that all human beings are born free and equal in dignity and rights regardless of race, colour, sex, language, religion, political or other opinion, national or social origin, property, birth or other status. The *Declaration* also includes a range of specified rights such as freedom of expression, political asylum, equal pay for equal work and the right to join a trade union.

As a concept, citizenship has broadened over time. Where once only the elite could be said to be citizens, now almost everyone is. Citizens may enjoy the services provided by the state (policing, judiciary, public services) in return for obeying the law. Clearly debate centres on entitlement to full citizenship. Some states have quite rigid citizenship laws which may be based on ethnic origin, or, more frequently, on some minimum residency requirement. Immigrants may find it difficult to obtain full citizenship rights. Such has been the case for Turkish people in Germany despite, in many instances, living in that country for many years and across a number of generations in some cases. Originally recruited as so-called *gastarbeiter* or guest workers, these Turkish workers and their families (including children actually born in Germany) were, until recently, denied full citizenship in the land that has become their home.

Despite its widespread applicability, it should be borne in mind that the conferring of citizenship does not in itself promote economic or social equality. Systems of privilege attached to private property or transmitted through the education system or other channels will still operate in a way which creates classes or layers within a society. Nominal membership does not automatically translate as equal membership in terms of access to power, and being residents within a particular territory does not automatically confer citizenship. Any discussion of citizenship raises wider issues of exclusion which needs to be seen in relation to specific groups of people and not just individuals (Storey 2003). Even where groups may have full citizenship, this may not be enough to prevent them from being treated as second-class citizens through various subtle and not-so-subtle forms of discrimination and victimization. Poverty and various forms of disadvantage serve to render many people less than first-class citizens in their own country. Minority groups such as people from certain ethnic backgrounds, gays or lesbians, people with disabilities, and women may often perceive their treatment as being different from the norm, thereby excluding them from supposedly 'normal' society (Sibley 1995). Although women obtained the right to vote in the UK early in the twentieth century (after a lengthy struggle – such rights are not automatically granted, they are fought for), more subtle forms of exclusion have continued to operate. Some groups may be the victims of collective antipathy. Gypsies or Roma people find themselves excluded from full citizenship in many countries. In some central and eastern European countries (and elsewhere), gypsies are treated as second-class citizens and persistently discriminated against in terms of employment, social services and in many other ways. Even though many countries, including the UK, have legislation that effectively makes it illegal to discriminate on these bases, it does not eradicate the racism, disabilism, sexism or homophobia which some people express, and which serves to make members of specific groups feel like second-class citizens. (These issues are returned to in Chapter 8.)

In a world of increasing mobility and international migration, the question of citizenship takes on added importance. Migrants may find themselves in vulnerable situations in many countries, denied the basic rights accorded to citizens. Refugees and people seeking political asylum are confronted with many problems related to their treatment in host countries. In Europe, a steady tightening of immigration controls has rendered it more and more difficult for many non-EU citizens to gain access to the bloc and it has become increasingly difficult for migrants, including refugees, from many 'Third World' countries to settle in Europe. Immigration policy is a key mechanism through which states can control access to their territory (Box 3.5). In the last decade, there has been an intense and shifting debate in the UK over ideas of citizenship. In part, this has been driven by concerns over immigration and the supposed need to identify a sense of Britishness within a multicultural society.

Box 3.5 Crossing borders

Migration is a long-standing social process but attitudes towards mobility are highly contingent on who is doing the moving. One of the peculiarities of migration is that the reactions to the migratory movements of some people are much more strident and controversial than others. As Hayter suggests 'migration for economic betterment, rather than being considered . . . a sign of enterprise and courage, is now regarded as criminal and shameful' (2000: 64) but Europeans and North Americans bettering themselves appears quite acceptable whereas people from Africa and Asia endeavouring to do so is another matter. During the colonial period, European settlers tended to act as though they had an automatic right not only to reside in faraway places, but also to control them. Many popular TV programmes in the UK promote the idea of people retiring to Spain or buying property in north Africa or eastern Europe. This is presented in a generally unproblematic way. However, the idea that people in Romania or Tunisia might make their way to Britain tends to be viewed somewhat differently. Those in richer countries have traditionally viewed their own migratory movements as a natural right while simultaneously restricting the movement of those from poorer areas. This highly unequal scenario is characterized by Bauman as one of tourists and vagabonds where 'the tourists travel because *they want to*; the vagabonds because *they have no other bearable choice*' (1998: 94). As Homi Bhabha memorably put it

> The globe shrinks for those who own it, (but) for the displaced or the dispossessed, the migrant or refugee, no distance is more awesome than the few feet across borders or frontiers.
> (cited in Gregory 2004: 257–258)

Contemporary neo-liberal economic orthodoxy has pressured poorer countries into opening up their economies to external penetration but simultaneously the inhabitants of those countries are faced with an array of shifting and often arbitrary rules when they try to enter the rich ones.

While traditionally citizenship has been concerned with relationships between people and their state, it might be argued that, in an increasingly inter-connected world, ideas of global citizenship need to be considered. This raises questions of people's responsibilities towards 'distant others', beyond the territorial confines of our own countries, and of our responsibilities to protecting the environment. The aid agency Oxfam, for example, has suggested that a global citizen might display an awareness of the wider world; respect and value diversity; have an understanding of how the world works economically, politically, socially, culturally, technologically and environmentally; be outraged by social injustice; participate in and contribute to the community; act to make the world a more equitable and sustainable place (www.oxfam.org.uk).

In focusing on the rights and duties which people have towards society, it is sometimes forgotten that people need to be empowered with the ability to actively participate in society. Some people may be unable to take full benefit of their rights, or they may be unaware of them. Equally, it may be difficult for some to participate fully in society as a result of poverty or lack of resources. People may nominally appear to have full citizenship rights, but may be denied economic power. They may, for example, have no control over their employer's decision to close down a factory, thereby depriving them of their right to employment and a decent income. Freedom of movement is of little value if you are too poor to travel. This also focuses attention on ideas of a more 'active' or insurgent citizenship, one that involves a questioning of government and the state rather than meekly submitting to actions done in our name or using only 'official' channels to make our views heard (Monbiot 2000; Painter and Jeffrey 2009). There is a need to move beyond a narrow conceptualization of citizenship and to see it in terms of a broader set of relationships (Hoffman 2004).

Summary

This chapter has introduced the concept of the state and has outlined various theories of the origins of states. It has cast light on the functions of the state and has proffered various perspectives on its role in capitalist societies. Finally, it has examined the relationship of the state to its citizens. The state is the world's most obvious territorial unit but it is a human creation resulting from the interplay of political processes. There is a need to be wary of not falling into what Agnew (1994) refers to as the 'territorial trap', whereby states are assumed to be immutable entities. States are historically contingent, they are dynamic and they reflect processes through which territory (and those living in that territory) is controlled. States do not just happen; they are created and reproduced, and occasionally subject to challenge. Through the appurtenances of the state we come to think in a territorial way. This state-centred approach reifies current political–territorial structures and tends to ignore the fact that, rather than being permanent features, they are historically contingent.

It is also important to bear in mind that states are more than mere spatial containers. It should be obvious that while states are spatially bounded entities, they are not static. States attempt to expand, to take over other territory, they colonize, they engage in conflict with other states. As suggested in Chapter 2, geographers have argued for the treatment of place as something more than an external 'commodity'. We live our lives in places and we imbue those places with meaning. Shared experience of places provides a basis for communal identity and a collective consciousness develops out of shared residency. Places are dynamic and people re-shape the places in which they live. In exactly the same way, states can be seen as dynamic entities, shaped and re-shaped through the inter-actions of human political activity so that they can be regarded as constantly evolving entities. One of the underlying elements in state-building, and one of the major sources of inter-state conflict, is the construction of nations, and it is to this that the next chapter turns.

Further reading

The suggested reading below contains material dealing with various facets of the state including some important seminal works. Also included are some useful contributions on the subject of political borders.

Anderson, M. (1996) *Frontiers. Territory and State Formation in the Modern World*, Cambridge: Polity Press.

Centre for International Borders Research, Queens University, Belfast: www.qub.ac.uk/research-centres/CentreforInternationalBordersResearch/

Fall, J. (2010) 'Artificial states? On the enduring geographical myth of natural borders', *Political Geography*, 29 (3): 140–156.

Jessop, B. (2003) *The Future of the Capitalist State*, Oxford: Polity Press,

Kabeer, N. (ed) (2005) *Inclusive Citizenship. Meanings and Expressions*, London: Zed Books.

Mann, M. (1984) 'The autonomous power of the state', *European Journal of Sociology* 25: 185–213.

Miliband, R. (1969) *The State in Capitalist Society*, London: Quartet.

Newman, D. and Paasi, A. (1998) 'Fences and neighbours in the postmodern world: boundary narratives in political geography', *Progress in Human Geography* 22 (2): 186–207.

Poulantzas, N. (1969) 'The problem of the capitalist state', *New Left Review* 58: 119–133.

Prescott, J. R.V. (1987) *Political Frontiers and Boundaries*, London: Unwin Hyman.

Rumley, D. and Minghi, J. (eds) (1991) *The Geography of Border Landscapes*, London: Routledge.

Sharma, A. and Gupta, A. (eds) (2006) *The Anthropology of the State. A Reader*, Oxford: Blackwell.

Smith, M. J. (2009) *Power and the State*, Basingstoke: Palgrave Macmillan.

4 Nations and nationalism

In the previous chapter, the nature and functions of the state, the primary building block in the political map of the world, were explored. This chapter examines the concepts of nation and nationalism, key elements in the creating and sustaining of a world of states. We live in a world where the existence of nations, like states, is taken for granted. 'Nation', 'nationality' and 'nationalism' are terms used regularly in the media and in everyday discourse. National identity is expressed through the singing of national anthems, support for national sports teams and in a variety of other, often mundane, ways. Nations and the associated political ideology of nationalism underpin the configuration of the world political map. Many of the disputes between countries and, indeed, many of those occurring within the borders of certain countries centre on competing nationalisms. In most instances, disputes between national groups are concerned with claims to territory. In many cases, these competing claims lead to extremely violent conflict, as in the former Yugoslavia, Northern Ireland and Kashmir.

Such conflicts make us aware of the more overt ways in which nationalism is asserted. However, as Michael Billig (1995) has argued, nationalism is an ever-present phenomenon and the nation is re-produced in many less obvious everyday ways. Differences can be observed in such things as forms of dress, cuisine (Indian, Chinese, Thai, Mexican), language and a variety of other means. These serve to reinforce our sense of a world comprised of nations so that 'the world of nations is the everyday world' (Billig 1995: 6). The flying of flags on public buildings and many other everyday phenomena may well go unnoticed and unremarked. However, through their very ordinariness, they inculcate a sense of national identity. This is what Billig refers to as 'banal nationalism'. He argues that in Western states nationalism, far from being a thing of the past, is in fact an endemic condition. The nation is constantly being flagged (often literally), thereby rendering it easier to mobilize the citizenry in its support during times of crisis. The term 'banal' implies something mild and relatively inoffensive but, as Jones and Merriman (2009) argue, the distinction between those overt or 'hot' expressions of nationalism and seemingly more banal ones may serve to mislead. Common practices such as street naming may reflect more deeply embedded versions of nationalism (Berg and Vuolteenaho 2009). What may seem ordinary or acceptable to one person may be oppressive or provocative to someone else. Monolingual English language road signs in Wales led to campaigns in the 1960s and 1970s to have Welsh names included. For ardent Welsh nationalists, mono-lingual signage was a signifier of cultural repression while for others the campaign was seen as irrelevant to their daily lives and concerns (Jones and Merriman 2009). The unremarked aspects of nationalism, those many and varied means through which the nation is constantly re-affirmed, serve as useful rehearsals for those moments when more overt forms of national mobilization are needed. Through banal affirmations, an unquestioning support can more readily be attained. This adds considerably to the relevance of Ernest Renan's observation in the late nineteenth century that the nation is a daily plebiscite whereby the willingness of the people to believe in its importance serves to ensure its existence (see Hobsbawm 1990).

This chapter discusses the origin of nations, leading to a discussion of national identity and a consideration of the territorial ideology of nationalism. Different forms of nationalism are outlined and the chapter ends with a consideration of the role and functions of nationalism and its implications.

Nation and state

To reiterate the point made in the previous chapter, nations and states are two distinct conceptual entities. The latter are agencies with power over citizens within demarcated territory while the former are more nebulous. Nations are social collectivities with an attachment to a certain territory. Hugh Seton-Watson suggests that 'a nation exists when a significant number of people in a community consider themselves to form a nation, or behave as if they formed one' (1977: 5). It follows from this that, to a large extent, a nation is more a mental construct than a physical reality. More concretely, Anthony D. Smith (1991) defines a nation as 'a named human population sharing an historic territory, common myths and historical memories' (1991: 14). Despite the differences between the two, nation and state are obviously closely interconnected and the relationship is succinctly described by David Miller: '"nation" must refer to a community of people with an *aspiration* to be politically self-determining, and state must refer to the set of political institutions that they may aspire to possess for themselves' (1997: 19, italics in original). The state can be seen as a political apparatus with the nation as its cultural corollary.

It is obvious that in some instances nations and states closely coincide. France is typically cited as an example of this – a state peopled mainly by those who regard themselves as French. However, such a view may have important implications for those perceived as 'non-French' but residing within the national territory. While France might be regarded as a close approximation to the idealized nation-state, where a territory is marked by a coincidence of national homogeneity and political control, there are many within France, particularly those of Algerian or Moroccan descent, who may suffer racial discrimination as a consequence of not being seen to properly belong to the nation. There are also those, such as Bretons and Basques, who may not necessarily see themselves as primarily French. The decision of the French government to ban the burqa and other Islamic face coverings in public in 2011 highlights the dissonance between the political apparatus of the state and the cultural characteristics and preferences of some of its citizens.

Despite much discussion of rights to national self-determination, there are many nations which do not have a state. One of the more notable examples is that of the Kurds, most of whom live in Iraq, Turkey and Syria, countries in which, by and large, they have been treated as second-class citizens. The aspiration towards an independent Kurdistan has not as yet been realized. Similarly, there are many states which encompass a number of nations, or peoples who would regard themselves as possessing a distinct national identity. The former Soviet Union and the former Yugoslavia (both of which imploded in the early 1990s) are good examples, embracing as they did a wide variety of national groupings. Another good example of the mismatch between state and nation is the United Kingdom, incorporating, as it does, the nations of England, Wales, Scotland and part of Ireland. (The peculiarities of nationalism within the UK will be returned to in Chapter 5.) Even when a nation has its own state, not all nationals reside within its borders. Leaving aside migration, the construction of state borders almost inevitably leaves some nationals living outside those lines. It is estimated that half of the Albanian 'nation' resided outside the boundaries of the Albanian state created in 1912, most of them in what is now Kosovo and western Macedonia (Hudson 2000) (Figure 4.1). Similarly many more Lao live in Thailand than actually live in Laos, a country in which only about half of the population might define themselves as ethnic Lao (Jerndal and Rigg 1998).

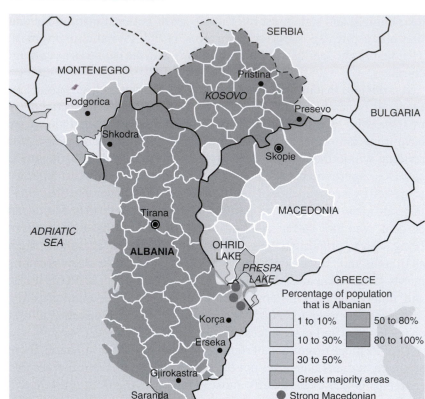

Figure 4.1 Areas with sizeable Albanian populations

There are also examples of nations which have formed a number of states. Reference is often made to the 'Arab nation', yet there are a number of Arab states such as Saudi Arabia, Bahrain, Kuwait and Egypt. Equally, it is obvious that many states contain minority national or ethnic groups. Examples include aborigines in Australia and 'native Americans' in the United States. Similarly, there are states where one or more groups wish to secede in order to create their own state to reflect what they see as their right to full nationhood. This is the case with Basque separatists in Spain, and Scottish and Welsh nationalists in the United Kingdom. The case of Northern Ireland is a specific form of secessionist nationalism in that those who wish to secede are seeking unification with, or incorporation into, another already existing state, the Republic of Ireland. This is known as irredentism. Particular issues exist in parts of the former Soviet Union where various groupings wish to secede from its successor states.

Chechnya fought wars of secession from Russia in the 1990s while Georgia has never really fully controlled all of the territory it claims as its own, such as the regions of Abkhazia and South Ossetia (Demetriou 2003).

Given the contemporary importance of nationalism and national identity and the strong emotions which it evokes, whether in outright war or in sporting or cultural contexts, it is important to have some understanding of the origins of nations and the evolution of nationalism as a political ideology.

The origins of nations

There is a tendency to think of nations as fixed, static entities which have always existed. For most people it is taken for granted that we are born into a national group which has a historic attachment to a particular territory. However, this is somewhat simplistic. How did this national group come into existence? How and why did people come to perceive themselves as part of a collectivity of this nature? The existence of nations is not as straightforward as it might at first appear and there are a variety of theories concerning the origins of nations. For purposes of clarity, these can be subsumed within two main strands of thinking. The first are referred to as primordialist or essentialist theories. The second are modernist or functionalist theories. It is important to bear in mind that the inevitable reduction of quite complex arguments into broad categories results in over-simplification. However, this should still present a flavour of the nature of the debate over nations and nationalism.

Primordialist theories

Primordialists see nations (as distinct from the more recent phenomenon of nationalism) as having quite deep-seated historical origins. Strict primordialists would argue that nations have always existed. This essentialist way of thinking sees nations as natural entities whose origins go back to time immemorial. Nations are seen as possessing some historical and immutable core. From this perspective, there is seen to be something which is essentially Dutch, or German, or Serb. These ethnonational identities might be seen as an extension of kinship which, in socio-biological terms, is seen as the 'natural' social unit. An alternative approach is that language, religion, race, ethnicity and territory (components in the nationalist mix) are basic organizing principles. Viewed in either terms, nations are seen as naturally occurring human phenomena. This extreme primordialist approach is reflected in the idea of a 'pure' race or nation. As an example, the words of an Irish folk song reflect a belief in something immutable at the core of the nation 'once upon a time there were Irish ways and Irish laws, villages of Irish blood waking to the morning'. Successive invasions of Ireland, first by the Vikings and then by the English, resulted in massacres and dispossession, yet '800 years we have been down, the secret of the water sound has kept the spirit of a man above the pain descending'[1].

Clearly there is a perception of something primordially Irish surviving all these waves of invasion, an immutable purity underpinning Irishness derived from a mythical pure past. In most instances, attachment to territory is part of this belief. Nationals may be deemed to have a deep bond with the territory of the nation, revealed through references to the national 'soil', the area of land seen to belong to the national 'imagined community'. In this way, fighting for, or even dying for, the land are seen as supreme acts of patriotism, ensuring that it does not fall into 'foreign' hands. In the late nineteenth and early twentieth centuries, notions of blood sacrifice in pursuit of national ideals were reflected in calls to defend the land (or national soil) and fight for it. Direct connections between blood and soil abound in these discourses. The Irish rebel leader Pádraig Pearse once wrote that 'the old heart of the earth needed to be warmed with the red wine of the battlefields' (cited in Townshend 2005). In a different context, later formulations of *blut und boden* (blood and soil) formed part of Nazi ideology and revealed a belief in an explicit link between territory and ethnicity. People fight (and die) for their country (or think that they do) indicating how strong associations with territory may be manipulated for the attainment of political ends. In the break-up of the former Yugoslavia, land and territory were central in a conflict characterized by attempts to purify portions of land of those seen as 'other' (Glenny 1999; White 2000; Mazower 2001). While it is easy to portray such events as irrational, for some people in specific contexts strong attachments to territory may appear to make perfect sense. In colonial (and former colonial) societies, for example, where land has been appropriated, a strong sense of ownership and defence may persist through succeeding generations. In places such as Ireland, Crowley argues that 'the collective memory of fighting for land ownership is a potent force' (2006: 135) and reflects a material (and not just sentimental) underpinning to an attachment to the land.

While this type of rhetoric may perform a useful function in rallying support for an oppressed population, it should be obvious that such views are absurd if taken in a literal sense and, in certain contexts, can have potentially dangerous consequences. They risk essentializing nationality and leading to witch-hunts against elements seen to dilute the essence of the nation. Such ideas (in their extreme forms) are often utilized by right-wing political groups. Hitler's assertion of Aryan racial supremacy was predicated on this primordialist idea. Contemporary neo-Nazi organizations such as *Le Front Nationale* in France and the British National Party in the United Kingdom have a version of this racialized thinking at the core of their political philosophy. These parties see their role as defending their nation against the dilution brought about through such processes as immigration or inter-ethnic marriage. Such extreme right-wing philosophies exist in a number of European countries, both east and west.

Ultimately the idea of the nation rests on those features which are assumed to make it distinctive and which link it back to antiquity. From this the nation can be 'celebrated' whether as a bastion of its people against its enemies or as a victim of oppression by others. This emphasis on distinctiveness helps explain the rush in Balkan countries in the 1990s, notably Croatia and Serbia, to emphasize the distinctiveness of their respective national languages despite the close similarities between them and that a common Serbo-Croat had been the official language of Yugoslavia. If nations rest on myth-making, it is also the case that their boundaries are determined in part at least by myth. Barth (cited in Kolstø 2005) has referred to 'diacritica': those boundary markers drawn from group culture which have a deep symbolism and significance and are therefore used to highlight the distinctiveness of a group. Furthermore, as Kolstø highlights, modern states have a tendency to lay claim to territory that may once have belonged to a supposedly historic predecessor state. This may lead to multiple claims on the same territory. For example, there are Serb, Greek, Albanian and Bulgarian claims to parts of the territory of the contemporary Macedonian state (Brunnbauer 2005), while Serb claims over Kosovo rest in part on an assumption that contemporary Serbia is a direct successor to pre-modern political formations.

A more nuanced and intellectually credible approach proffers a theory of national origins which accepts their constructed nature, but yet sees them as having a long historical lineage, though rejecting the idea that they are 'natural' entities. Anthony D. Smith (1986), while rejecting essentialist ideas of the primacy of nations of the kind outlined above, nevertheless sees nations as originating from what he terms *ethnies*, which were extensions of kin-groups. These groupings developed feelings of community over time. Feelings of belonging together, of collective identity and of attachment to their locality would have evolved gradually. Smith argues that such feelings have waxed and waned with some ethnies becoming more self-conscious and dominant than others. Vertical suppression of subordinate ethnies would have taken place as the intelligentsia played a role in inculcating a common culture aided by processes such as the harmonization of vernacular languages. This would have led to the territorial diffusion of cultural norms which then became dominant. This contrasts with an alternative mode of development, termed by Smith 'lateral amalgamation'; in this instance, there is a merging of ethnies in which inter-ethnie differences become submerged into a new common identity. In this latter case, the nation could be said to come into being through a process of bureaucratic incorporation whereby localities on the periphery are subsumed into a broader 'national' polity. In essence however, nations are seen to form around ethnic cores. The reconstruction of these and their relationship to a specific territory gives rise to the nation. For Smith, nations are not natural entities but they are political–territorial phenomena in which ethno-national symbols based on historical events and characters assume huge significance.

Modernist theories

Rather than seeing nations as extending back through time, some suggest that they are relatively modern creations, 'invented' as recently as the eighteenth and nineteenth centuries, although various 'proto-nationalisms' might be recognized prior to that period. The argument here is that nations served, and continue to serve, a particular purpose. They provided a mechanism whereby modernizing capitalist societies could be ordered. Hence, these theories can be termed instrumentalist or functionalist. Rather than being viewed as having deep-seated historical origins, the nation is seen as a construct devised to serve specific purposes. Modernist theorists such as Eric Hobsbawm and Ernest Gellner argue that nations and nationalism reflect the needs of a particular economic and political configuration. In this way, they arose as a response to the functional needs of an industrial capitalist system rather than the historic inevitability of communal feelings.

During different historical epochs, people owed loyalty to god, to a monarch or to an overlord. In the modern period, dynasty and deity have been replaced by the nation. Whereas people may have fought for their religion (clearly some still do), now they fight and die for their country, for their nation's right to self-determination, or for its right to defend its territory. Industrialization is seen as the catalyst in this change. It is argued that nationalism only arose in industrialized societies, not in predominantly agrarian ones. Gellner (1983) argues that the new social divisions of labour associated with industrial societies needed culturally homogenous literate populations and the idea of the nation emerged as a result of this need. In order to achieve the political support of the masses, it was necessary to create a political infrastructure. This emerged as nationality, leading to feelings of national identity which served the needs of the industrial age through the creation of overarching feelings of broad community. Clearly this was a lengthy process and not something which happened overnight and, while many would accept that nations are constructions and relatively recent ones, intense debate still surrounds the 'moment' when a specific nation could be said to have come into being (Kumar 2003).

An important contribution to the modernist approach on the spread of nationalism has been provided by Benedict Anderson. He suggests that nations can be seen as 'imagined communities' (1991). They are imagined primarily in the sense that they provide feelings of belonging, solidarity and commonality among people who have never met and, in most cases, never will. The harmonization of vernacular languages, a process facilitated by the spread of printing and publishing (what Anderson refers to as 'print capitalism'), played a key role in engendering shared feelings. Allied to the rise of literacy, there was the gradual recognition of a collectivity that shared a language. As Anderson describes it, there were people who could read a book knowing that many others elsewhere were doing the same thing. These people could thus be said to be part of an 'imagined community'. Of course, just as vernacular literature united people, it also excluded others. A contemporary way of thinking about this is football supporters cheering on their national team in a game, whether in the stadium or watching on television. The event could be said to make 'real' the otherwise imagined national community as people throughout the country realize that others are cheering (or despairing!) just as they are. What has happened is something of a scalar shift with people moving from highly localized senses of identity to a much broader sense of the 'country' as a territorial entity with which to identify (Billig 1995; Etherington 2010).

It follows from this that nationalism is, in the words of Gellner, 'neither universal and necessary nor contingent and accidental' (1997: 10). Certainly the word 'nationalism' does not appear to have entered into common usage until the nineteenth century and the French revolution is seen by many historians as the catalyst in bringing about a territorially based concept of nationhood. In any event, unlike extreme primordialist ideas, nations and nationalism are seen as political products, not natural characteristics. The debate between the primordialist school and the modernist school, as mediated through Smith and Gellner, is in part a debate over emphasis. From the modernist perspective, nationalism is a force which can utilize pre-existing cultures or, in many instances, completely obliterate them. Thus, nationalism is not the awakening and assertion of these mythical, supposedly natural given units. It is, on the contrary, the crystallization of new units, suitable for the conditions now prevailing, though admittedly using as their raw material the cultural, historical and other inheritances from the pre-nationalist world.

(Gellner 1983: 49)

In stressing the 'newness' of nationalism, Smith (1986) feels that modernists underplay the role of ethnic origins and the importance of nationalist myths associated with those origins. As Perica (2005) suggests 'new' nations utilize old ethnic traditions in order to highlight the continuity between the modern nation and its ethnic precursor. Modernists do not necessarily dismiss these myths but they see them as masking more 'real' social forces, rather than viewing them as 'independent' entities.

For those who wish to generate or solidify a sense of the nation, the construction of a national past is clearly important in maintaining a sense of nationhood. This construction results from the efforts of nationalist ideologues keen to forge a heroic vision of their nation. Instrumentalists tend to view such ideologues promulgating visions of past national glory as catalysts in present political and national debates utilizing myths of the past to forge a different future. A primordialist view is that while intellectuals clearly do this, they can only do so by virtue of the strength of the myths in the first place. In other words, rather than 'inventing' a national past, the essence of the nation allows the intellectual propaganda to flourish.

As well as a national past, a national geography is created and reproduced and the construction of national maps has played a key role here (Hooson 1994; Etherington 2010). The political geographer Jim MacLaughlin (2001) writes of his father going to school in the 1920s in the then recently created Irish Free State (now the Republic of Ireland) where the children traced a map of Ireland onto paper from an outline of the island hammered into the school room floor with nails. Subsequently, Irish school exercise books had maps of the country on the outside cover serving as an everyday reminder of the shape of the nation. These examples and others such as the hanging of political maps on schoolroom walls and other related phenomena serve the purpose of familiarizing people with the territory of their nation. They also reflect the huge importance of the education system in inculcating a sense of national awareness and culture (Kumar 2003). Through such activities, the country is made real and tangible leading to what might be seen as a spontaneous identification with it. In Turkey, maps have been a key instrument in the cultural reproduction of the nation (as elsewhere). The use of a national map of Turkey as a type of commonly reproduced logo familiarizes people with the shape and extent of the nation's territory; the map and the logo can be seen as mundane signifiers (Batuman 2010). The flag of the recently independent Kosovo is a map of the country!

It should be clear from the above that there is considerable debate over what precisely a nation is and profound arguments over its origins. What cannot be disputed, however, is the contemporary importance of nationality in influencing the ways in which we view territory and our sense of identification with that territory.

National identity

As has been indicated, the nation refers to a group of people who share particular historical–cultural characteristics or imagine themselves to do so. Nationality refers to the condition of belonging to a nation. At its most basic, nationality can be seen as a mechanism of social classification. People know who they are and who others are. We are accustomed to seeing a person's national identity inscribed adjectivally. Thus, Carlos Tevez is a *Argentinian* footballer, Youssou N'Dour is a *Senegalese* singer, Ismail Kadare is an *Albanian* writer. There are two components of national identity, according to Verdery (1996). The first is a collective identity which refers to national characteristics and so-called national traits and may include such features as language and style of dress. This is an identity which is shared by the members of the national community. The second meaning to national identity is the individual member's sense of self as a national. An individual's feeling and self-identification as 'English', 'French', 'Spanish' is an important component in their self-perception. It refers to a feeling of belonging to a nation.

In many instances, people's national identity may be officially defined in terms of where they were born. For many, this may not be a problem; however, people may often define themselves in different terms. Many second-generation Irish people living in Britain may see themselves as Irish and many among the north African diaspora in France may see themselves as Algerian, Moroccan or Tunisian rather than French. Throughout eastern Europe there are numerous nationalities resident within the borders of other states. Thus, there are many ethnic Russians living in the Baltic states of Estonia, Latvia and Lithuania and in the other former member states of the USSR, referred to in Chapter 3. Many ethnic Germans 'returned' to Germany from Russia following the collapse of communism, returning to a country in which they had never lived. It follows that national identity is not simply a function of where a person is born.

Guibernau (1996) sees national identity as composed of five key elements:

1   Psychological – consciousness of forming a community.
2   Cultural – sharing a common culture.
3   Territorial – attachment to a clearly demarcated territory.
4   Historical – possessing a common past.
5   Political – claiming the right to rule itself.

Obviously these five characteristics are closely interlinked. Within this milieu elements such as language, religion and social mores may take on particular significance. Many nations are seen to possess their own language, while in some the majority of members adhere to a particular religion. In these cases, language or religion may be the key defining characteristic of the nation. For some, language is often deemed to be the primary binding mechanism of the nation. The Welsh language is seen in some strands of Welsh nationalism as an integral part of Welsh identity. However, many Welsh people view the language issue as exclusionary. Throughout much of industrial south Wales, many people feel no affinity with the Welsh speakers of the predominantly rural areas of north and mid-Wales and some may feel they are viewed as being less Welsh as a consequence of their inability to speak the language (Figure 4.2). More broadly, though it may be held up as the primordial glue binding together the nation, a national language is usually an outcome of the creation of the nation rather than its precursor. Harmonization of vernacular languages, or prioritizing of one form over others, is usually part of the nation-building process and it has been argued that 'cultural characteristics such as national languages, national religious belief systems and national administrative networks were as much the product as the cause of nineteenth century European nationalism' (MacLaughlin 1986: 324).

The various elements of a national identity need to become known and assume meaning among a populace in order for them to become binding influences. Miroslav Hroch (1993) has suggested that there are five key elements in considering how a nation comes to be. First, it needs a past which provides a national history. Second, culture works to provide a common ethnicity, language, religion, and so on. Third, modernization, associated with industrialization and the development of capitalism, allows for the creation of political systems with a centralized administration (the state) while the fourth element, communications systems, facilitate the transmission of ideas through language, education, the media and so on. Finally, agitation through the invention and promotion

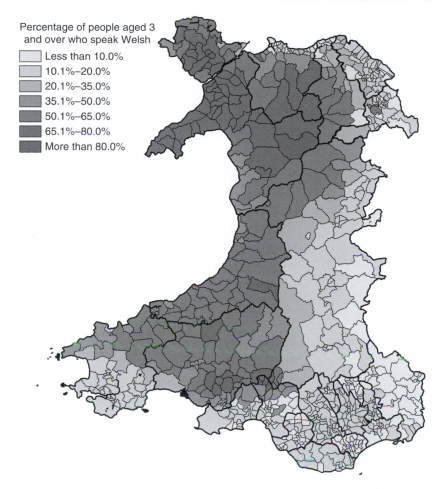

Figure 4.2 Welsh language speakers

of symbols, festivals, events, etc., engenders an emotional attachment to the idea of the nation (Figure 4.3). Ultimately, it is apparent that 'a nation presupposes a relatively high degree of cohesion and common consciousness among its members, together with a regular and effective system of communication between them' (Kumar 2003: 47). In order to constitute a nation, people must have a sense of themselves as part of that nation. As Muir (1997) points out, national identity is important precisely because people feel it to be important. While it can be argued that there are no objective criteria by which national identity can be measured, its subjective component – the extent to which people believe in it – is very important indeed. A nation is clearly more a mental construct than a concrete reality.

Figure 4.3 Celebrating Hungary's national day, Budapest (Source: Author)

While most countries contain an array of ethnic groupings, not all will necessarily develop nationalist aspirations, but given the historically constructed nature of national identity, it follows that any nation will contain within it the seeds of other nations. In this way, separatist movements arise, such as the Basques in Spain, the Albanians of Kosovo in Serbia, Tamils in Sri Lanka. Even newly formed states such as Azerbaijan and Georgia find themselves faced with secessionist claims, as indicated earlier. If French-speaking Quebec attained independence from Canada, it is quite probable that it would be confronted with claims for Iroquois independence.

It is easy to fall into the trap of thinking that a national culture is clearly identifiable and that this presents grounds for the existence of a distinctive nation. While some geographers in the past attempted to identify cultural heartlands and domains, the idea that we can identify clear cultural boundaries is simply not tenable. A look through a contemporary atlas of art, for instance, indicates a series of maps of what could in many ways be seen as culture regions but in which the boundaries depicted do not accord with present-day political borders (Onians 2004). While identities are often ascribed to people, they themselves may have a more complex perception. Many Indian people in Britain may feel a sense of both Britishness (their country of birth) and Indian-ness (the country of their parents' or grandparents' birth). For some, there may be a sense of cultural hybridity rather than a neat acceptance of the straitjacket of a singular identity. Ideas of hybrid identities have been common in the United States and are likely to persist, even to grow, in an era of improved communications, increased migration and other globalizing processes (Box 4.1). While some seek to pigeonhole people in terms of their identities, this is to ignore the complexities of people's biographies and family histories. Some argue that people should choose one national identity above others. A controversy in France in 2011 over the selection of players from non-French ethnic backgrounds for the national football team engendered debate over the extent of affinity felt by such players towards both France and the team and a questioning of the loyalties of those possessing dual or multiple national identities. While some may see a definitive national identity as 'natural', Kumar (2003) argues that the insistence on one singular national identity is the anomaly and that the common experience of many is of composite or multilayered identities. The construction of Argentinian and Chilean nations and the solidifying of difference (consequent on Spanish colonization), for example, has implications for indigenous populations within those countries and particularly for many living in the regions close to the border between them (Escolar 2001).

Box 4.1 Sport and national identity

The complexity of national identity can be illustrated through reference to sport, where there is an increasing tendency for international sports people to represent a country other than the one where they were born or grew up. For example, the South African-born cricketers Kevin Pietersen and Jonathan Trott represent England. Often (though not always) this is a decision to represent the country of their parents or grandparents birth. In football from the 1980s onwards, the Republic of Ireland football team has included sizeable proportions of non-Irish-born players, the sons and grandsons of Irish emigrants (Holmes and Storey 2011). Similar tendencies are clear in the selection of English-born players for teams such as Jamaica and Trinidad and Tobago, while similar colonial connections explain France's selection of players born in African countries but brought up in France, such as Marcel Desailly and Patrick Vieira (born in Ghana and Senegal respectively). Conversely, some African countries have begun increasingly to select French-born players of African origin thereby reclaiming some of the sons of their extensive diasporas. The Algerian football team at the 2010 football World Cup finals contained a sizeable number of players who had been born in Europe, mainly France. In a newspaper article in 2011, the footballers Benoit Assou-Ekotta and Sebastian Bassong spoke of their identification as Cameroonian rather than French. Both were born and grew up in France but have elected to play international football for Cameroon, the country from which their parents came.

Obviously, national identity is by no means the only identity to which people lay claim. Nevertheless, it is an important one in a set of over-lapping identities. Thus, people may be male, female, heterosexual, homosexual, black, white, etc. Alongside these various identities, people also have a national identity. Which of our many identities predominates will depend on the context. An individual's sense of national identity may be heightened at particular moments. For example, during times of international crisis, people's sense of national identity may become more prominent with references to supporting 'our' soldiers as they fight for 'us'. In a less serious context, sporting events, or even occasions such as the much-derided Eurovision Song Contest, may spark heightened feel-ings of national identity, manifested through support for the represen-tative(s) of the nation.

It is not uncommon for feelings of national identity, otherwise dormant, to come to the fore when visiting (or working or living in) another country. This may, in part, be a consequence of the 'native' population's sense of the 'visitor' as 'different' and in part attributable to the visitor's sense of this difference. Feelings of missing the home country may of course be linked to a sense of missing the comfort and re-assurance of familiar surroundings, familiar faces and places, and dealing with 'strange' customs, a foreign language and so on. These feelings of familiarity and alienation are linked, therefore, to a person's sense of place, as discussed in Chapter 2. It follows from this that we should be mindful of the complexities faced by international migrants trying to live their lives in places and negotiate their cultural identities and connections (with other places) within a world that is often far from welcoming or tolerant (Gupta and Omoniyi 2007; Antonsich 2009).

National or ethnic identity can be seen as relational: it can be seen as contingent to 'others' who are seen as possessing a different identity (Box 4.2). As Edward Said (1995) argued in his analysis of orientalism, the 'other' becomes objectified. It then follows that our identity is defined in terms of difference from this objectified other. Given the somewhat ephemeral nature of national identity, it is sometimes argued that it is easier to define it in terms of who one *is not* as opposed to who one *is*. Being Welsh may be most easily articulated in terms of *not* being English; to be Portuguese may be conveyed in terms of *not* being Spanish. During the colonial era, 'Englishness' and 'Frenchness' could be seen in terms of supposedly 'civilized' traits not possessed by those 'others' being colonized. This process allows the members of one nation to view themselves as superior to those of another, thus legitimizing anything from casual disdain, through discrimination, to genocide.

Box 4.2 Thai national identity

Winichakul (1996) raises the question of what constitutes 'Thainess' in the construction of Thai nationality. What distinguishes Thais from Burmese, Cambodians or Laotians? In part, the answer lies in the contact between indigenous elites and Europeans. In pre-nineteenth-century Siam (the previous name for Thailand), the concept of rigidly demarcated borders was unknown and the idea of overlapping sovereignty was generally accepted and understood within 'frontier' zones. However, Siam found itself in increasing contact with European powers. Specifically, France controlled territory to the east in what is present-day Cambodia, Laos and Vietnam, while the British controlled Burma to the west (Figure 4.4). Relations between the Europeans and those in power in Siam gradually required a firming up of territorial boundaries which had previously been somewhat ephemeral. In order for these boundaries to have meaning on the ground for local people, and to thereby further the 'construction' of the state, a sense of nationhood needed to be created. From the late nineteenth century onwards there was a need to move towards inculcating a sense of national identity.

In defining 'Thainess' something called 'un-Thainess' also needed to be identified. The former would thus be the opposite of the latter and would embody the positive traits absent in the latter. As Winichakul demonstrates, Thai national identity arises as a relational notion, an identity defined in terms of what is it *not* as much as what it *is*.

The domain of Thainess is also defined by what is 'non Thai'. Once the un-Thainess can be identified, its opposite – Thainess

> – is apparent. According to a Thai historical perspective, for
> example, the Burmese were aggressive, expansionist and
> bellicose, while the Khmer were rather cowardly but oppor-
> tunistic. The Thai, unsurprisingly, are taken to be the mirror
> image of these traits – a peaceful, non-aggressive but brave and
> freedom-loving people, precisely the description in the Thai
> national anthem. The existence of un-Thainess is as necessary
> as the positive definition of Thainess.
>
> (Winichakul 1996: 219)
>
> This highlights the idea of defining nationality in negative terms:
> in seeing it quite explicitly as relational juxtaposed to other
> 'inferior' national identities. It also, of course, demonstrates the
> artificiality of national distinctions for 'what has been believed to
> be a nation's essence, a justifiable identity, could suddenly turn out
> to be fabrication' (Winichakul 1996: 220).

It is also important to note that national identity is shaped, in part at
least, by that nation's sense of its own role in the world. Thus, the perception of US identity is heavily influenced by the perceived importance
of the country in international relations. Many may feel proud of their
American identity because of what they see as the important role of the
USA in world politics. Others may feel ashamed of their American
identity because of aspects of US foreign policy. By the same token, many
British people take pride in their country by virtue of what they see as
its former greatness. Britain's imperial past and its economic pre-
eminence during the era of industrialization, when it was seen to lead
the world, may be a source of great national pride to many. For others
though, this same history may give rise to a sense of shame over acts
committed in the name of the British people and the subjugation of the
indigenous inhabitants of former colonies. In this way, geopolitics and
geopolitical relationships interact with national identity (Dijkink 1996).

In the 1930s, Stalin parcelled the Soviet 'Republics' of the USSR
around what he saw as 'national' criteria based on language, culture and
territory in order to resolve nationalist tensions, thereby furthering the
communist project. The boundaries imposed around the Soviet 'nations'
resulted in its successor states containing many national minorities. This
serves to demonstrate both the difficulty and the illogicality of defining
nationality in precise terms and also the self-defeating and ultimately

Figure 4.4 Thailand

counter-productive nature of the exercise in the first place. The creation of Uzbek, Kazakh, Kirghiz, Tajik and Turkmen 'republics' was a Stalinist theoretical construct, not a reflection of primordial national groupings (G. Smith 1995). However, once constructed, these territorial entities subsequently engendered a sense of nationality among the populace, though with enduring ethno-national tensions in places such as South Ossetia, Chechnya and the Fergana Valley, as noted earlier.

In summary it can be said that national identity is important to many people in providing them both with a sense of 'self' and with a feeling of being part of a larger collective. This identity is, of course, a 'felt' one rather than an 'objective' one. It is also an identity that may be defined oppositionally in terms of what one is not rather than what one is. Many people may possess multiple national identities rather than a single unambiguous one. While national identity can be seen, in some senses, as defying 'rational' categorization, the construction of national categories creates and reinforces a very real sense of identity.

Nationalism

As already observed, many people identify strongly with the nation to which they feel they belong. At its most extreme, this may be apparent through a willingness to fight or even die for one's country. In more peaceful circumstances, this sense of identification may be manifested through such activities as support for national sporting teams or individuals and through singing, or observing solemnity during the playing of, the national anthem. Even at the comparatively innocuous level of self-definition as 'Scottish', 'French', 'Vietnamese', etc., people almost unthinkingly see themselves in terms of their nationality. This sense of identification is seen to reflect an ideology of nationalism. It is an ideology in the sense that it encapsulates a set of beliefs and practices, a world view, which people come to accept as 'natural'.

The communal sense of identity that is at the heart of nationalist ideology gives rise to a sense of a 'national will' which unifies all members of the nation. This reification of the nation presents it as a 'natural' and largely unchanging phenomenon. As will be seen, this is very far from being the case. In this way, international football supporters talk of 'their' victories, 'their' defeats, etc. For some supporters, defeat for their international team is a defeat for them. It is they who have lost, not merely the players on the pitch. Similarly, a desire to attack another nation can be put into practice through an assault on its people.

It is important to observe that, as noted earlier, the nation refers to a social collectivity rather than territory. Nevertheless, it is a territorial concept in that the group of people concerned feel an attachment to a particular territory and national disputes invariably centre on struggles over the control of land. Because of this, nationalism is a territorial ideology reflecting an affinity to a particular space and one which, in its more 'active' form, seeks to maintain or to attain political independence (and in some cases dominance) for the nation and, hence, for its territory. Anthony D. Smith defines nationalism as 'an ideological movement for attaining and maintaining autonomy, unity and identity on behalf of a population deemed by some of its members to constitute an actual or potential "nation"' (1991: 73). This reminds us that a sense of nationhood, while it may assume an 'essential' quality, is ultimately a tool useful in the attainment of particular political goals.

National secession

Secessionist nationalism is where national groups wish to see the creation of alternative political units carved out of part of the territory of the existing state. Classic examples of this are the struggle for Basque separatism which seeks a Basque state wholly independent from Spain (see Box 4.3). The Basque conflict led to numerous military attacks in Madrid and elsewhere and assassinations of Spanish politicians in the latter part of the twentieth century. Northern Ireland is another classic example of separatist nationalism. Nationalists within the north of Ireland wish to see the region become part of a united Ireland. This irredentist claim is opposed by unionists who see themselves as British and wish to remain so (see Chapter 5). While the Basque and Irish situations have produced militant conflict for the Spanish and UK states, milder sub-state nationalisms exist in both countries. In the case of Spain, significant Catalan and Galician minorities exist in the north-east and north-west respectively, while in the UK debates continue over full Scottish and Welsh independence.

Box 4.3 Basque separatism

The region referred to as the Basque country is part of northern Spain extending northwards into southern France (Figure 4.5). It is an area with a distinct culture, including its own language. Historically, the region has enjoyed periods of relative autonomy but it has never had complete political independence and resentment has been continuous as the region's identity has been devalued within the process of Spanish state-building. Calls for a separate Basque state arose in the early part of the twentieth century when Franco's fascist regime opposed any autonomy for the region and was distinctly hostile to Basque culture and language. The Basque language, like Catalan, had been effectively banished with Castilian promoted as the 'national' language of Spain. This overt oppression helped bolster support for militant nationalism and saw the rise of *Euskadi ta Askatasuna* (ETA – Basque Homeland and Liberty), formed in 1957 out of a variety of pre-existing factions, as a military organization aimed at secession from the Spanish state. Prior to an ongoing cessation in hostilities, ETA was involved in many military attacks in Spain and many have died in the ensuing conflict.

When Franco's reign came to an end in the 1970s, Spain moved towards a federal solution to the problem of regional instability and granted the Basque country a degree of autonomy, within a federated Spanish state. Thus, the Basque regional government administers its own territory and even has its own police force. Nevertheless, control over matters such as foreign policy and general economic planning resides with the central government in Madrid. Despite this, many still see complete independence as the ultimate goal. There is now an evolving peace process and ETA has declared a cessation of military activities. Walls in the towns and cities of the Basque country continue to be emblazoned with graffiti asserting the right to independence.

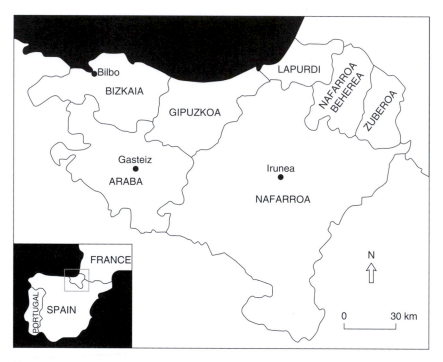

Figure 4.5 Basque country

Within Africa there are numerous examples of ethnic differences giving rise to secessionist tendencies. Eritrea's protracted war of secession is a prime example. While Eritrea succeeded in its secessionist struggle, failed attempts abound. Examples include the Ibo insurrection within Nigeria leading to the Biafra War of 1967–70. Part of the problem for African states is that they were created by external colonizing powers with the result that the concept of the state and its present configuration may have little effective meaning for its citizens, despite nation-building attempts by political elites. Other forms of loyalty exist and ethnic divisions may render the state inherently unstable. Conflicts within Democratic Republic of Congo, Nigeria, Ivory Coast and elsewhere reflect a deep unease among some with those states as presently constituted.

Within the complex mix of sub-state ethnic and nationalist tensions, cultural differences may assume huge importance. In this case, elements such as language may act as a centrifugal force causing instability. Belgium is an example of a country with a significant language divide between a French-speaking Wallonia and a Flemish-speaking Flanders. Since the 1960s the country has followed a strict language equality policy which also includes a degree of recognition for German, spoken in areas along the German border. The recognition of the language divide was cemented with the adoption of a federal structure in 1993 resulting in a tripartite territorial division between Flanders, Wallonia and Brussels which, as the capital city, is officially bi-lingual although it is located within Wallonia (Figure 4.6). Flemish is the official language of Flanders, French is the official language of Wallonia. Tensions have arisen periodically which strongly suggest that identification with language zone is, perhaps, stronger than identification with Belgium. Under these circumstances, some feel that there is no sufficiently strong underpinning for the Belgian state. Murphy (1993) has pointed to the limited interaction between residents of the two main regions and has also highlighted the fact that, once regional boundaries are firmed up, as they were in the 1960s in the case of Belgium's linguistic regions, they tend to reduce interaction across those boundaries. Just as national boundaries serve to create difference, so too do internal borders. As suggested in Chapter 3, boundary creation gives rise to its own dynamic whereby previous divisions become institutionalized and enshrined as part of everyday existence. The formalizing of ethno-cultural divisions solidifies identities rather than transcends them. In this way, people's everyday lives become territorialized as a consequence of the political–territorial framework within which they live. The example of Belgium seems to suggest that differences may become more deeply inscribed, and subsequently reinforced, through territorial divisions. In this way, those divisions which are the reason for federalism in the first place become reified and reproduced as a consequence of federalism (Agnew 1995). A sense of place and identity is reinforced through the federal structure further highlighting the way in which politics and territorial identity are inter-meshed.

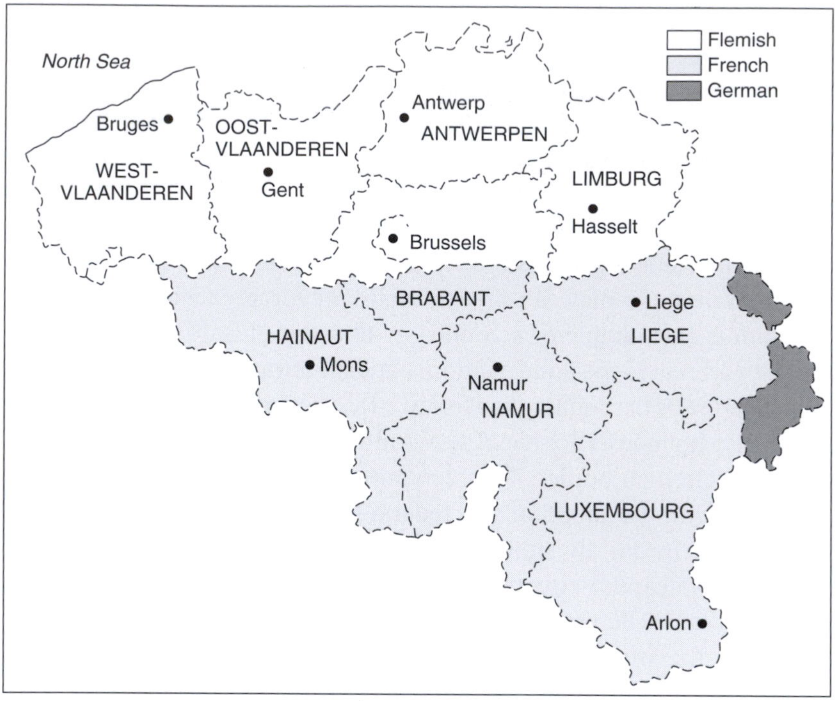

North Sea

Flemish
French
German

Bruges •

OOST-
VLAANDEREN

WEST-
VLAANDEREN

Gent •

Antwerp •
ANTWERPEN

LIMBURG

Hasselt •

• Brussels

BRABANT

• Liege
LIEGE

HAINAUT

• Mons

Namur •
NAMUR

LUXEMBOURG

Arlon •

Figure 4.6 Belgium's linguistic divisions

In Canada, a language issue also threatens the continuance of the Canadian state. In Quebec, there is a French-speaking majority and Canada has adopted a bilingual policy in order to satisfy the demands of those in Quebec who feel that for far too long Canada was an anglophone state which essentially denigrated 'French Canada' and Quebec's distinctiveness. French enjoys a linguistic equality throughout sizeable parts of Canada where it was never widely spoken. In the early 1990s, a referendum on independence was narrowly defeated, with roughly 55 per cent opposing a break-up. Nevertheless, a distinct divide between 'French Canada' and 'English Canada' continues to pervade political and cultural life. Indeed, the Quebec case is only one of a range of potential secessionist issues facing the Canadian state. In 1999, a self-governing Inuit territory was created in the north, while throughout Canada indigenous First Nations land claims represent a territorial challenge to state hegemony (Figure 4.7).

Socio-economic differences within countries have been seen as a consequence of the development of cores and peripheries with some regions remaining marginalized while others are seen to be economically dominant – a situation likened to a form of 'internal colonialism' (Hechter

Figure 4.7 'First Nations' heritage site in Vancouver, Canada (much of the land in the city is claimed by First Nations groups) (Source: Author)

1975). These core–periphery tensions can also place pressure on the state with calls for secession emanating from either the periphery, or, perhaps less frequently, the core. High levels of disaffection in particular regions might lead to calls for secession or for greater autonomy. Peripheral nationalisms may have as much to do with regional inequalities as with any innate sense of nationhood. Peripheral places are dependent on cores and a sense of resentment may develop towards an 'imperial' centre which is seen to exert control over them. In the UK, moves at asserting Cornish identity are strongly related to ideas that the county is being peripheralized with the disappearance of its tin mining industry and an increased reliance on tourism. It is argued by some that Welsh economic peripherality has strongly influenced Welsh nationalism and has thus bolstered a sense of Welsh identity. Marginalization of Flanders has in the past been a mobilizing force in assertions of Flemish identity, thereby contributing to the territorial divide within Belgium. A different type of periphery dissatisfaction occurs in Spain where the apparent prosperity of the Basque country and Catalonia has, in the past, been seen as a reason for separatism: in order to avoid these riches being siphoned off by the Spanish state. In Italy, the Lombard League seeks independence for northern Italy, historically the more industrialized and prosperous part of the country, and separation from the relatively poorer and more agricultural rural south.

Forms of nationalism

There are many different classifications of nationalism proposed by various authors. One division revolves around a distinction between a civic–territorial form and an ethnic–genealogical one (Kohn 1967; Plamenatz 1976). In many countries, most notably within western Europe, national identity is ascribed on the basis of where one is born. Thus, if a person is born in Britain, they are rendered British. Of course, a person may also acquire British or US nationality through the fulfilment of criteria based on behaviour (residency for a certain period of time, language proficiency, etc.). This is seen to contrast with many eastern European countries and with Germany where national identity is based on ancestry. Attempts to resolve conflict in the countries of the former Yugoslavia centred on complex questions of national or ethnic identity which are not easily resolvable through reference to place of birth. In Germany, citizenship is available to those of German origin even though their families may have lived elsewhere for centuries. In the 1990s, many ethnic Germans from Russia and Poland moved to Germany (following the collapse of communism) and were accorded German citizenship despite having never lived in the country. By way of contrast, ethnic Turks who have lived for many years in Germany (as a consequence of encouraging the in-migration of *gasterbeiter* or guestworkers), or have even been born there, could not until recently obtain German nationality.

This distinction between an ethnic (eastern) and a civic (western) nationalism carries certain connotations with the ethnic variant being seen, perhaps understandably, as less progressive than the civic version. The latter is seen as furthering the cause of democracy while, in contrast, the other is seen as 'negative', being based, it is argued, on emotion. As a result, ethnic nationalism is deemed irrational, unlike the civic-based nationalisms of the west, of which Britain, France and the United States are seen as the exemplars. As a consequence of this, there is a tendency to frown upon the apparently tribal-like nationalisms of the Balkans and elsewhere in eastern Europe. This is in contrast to the supposedly fully evolved civic nationalism of western Europe and North America. However, many Western countries might do well to reflect on the existence of ethnic tensions and overt racial discrimination within their own borders. While this has not necessarily led to ethnic cleansing and mass murder, it does, nevertheless, suggest that the civic–territorial nationalisms have been less than fully successful at integrating all members of society into a common and egalitarian polity. Certainly, many people of West Indian or south Asian origin in the United Kingdom and many Algerians and Moroccans in France might take a somewhat more jaundiced view. It can also be argued that civic nationalisms themselves are built around particular 'ethnic myths'. For example, notions of British tolerance and a 'stiff upper lip' are often presented as 'superior' traits to the apparent ethnic or 'tribal' myths which are seen to generate conflict. This ignores the constructed nature of these supposedly national traits. Moreover, even in nations claiming a civic basis for membership, there still exists an (presumed) ethnic component that links the people to that land and to that specific geographic space (Etherington 2010).

Another basic distinction within the literature on nationalism is that between forms of 'top-down' nationalism on the one hand and 'bottom-up' nationalism on the other. The former occurs where a political elite attempts to construct a nationalism which inculcates loyalty to the state. In these examples, the creation of the state pre-dated a sense of national consciousness. In this way, nationalism as an ideology follows state formation rather than being a pre-condition for it. England and France are cited as examples of this process, referred to by Tom Nairn (1977) as national determinism or bourgeois nationalism. In order for this to occur, it requires the invention of a nationalist ideology and the identification of key traits which serve to bind the nation together. In this way, the inculcation of national identity serves a clear political end, namely the maintenance of state hegemony.

Alternatively, we can identify national self-determination or nationalism from below. This is a nationalism born of a desire to shake off a particular imposed rule. It is seen to emanate from the 'periphery', quite often applied to Third World countries, but with European examples such as Basque nationalism and that of the 'Celtic fringe'. This is a nationalism generally constructed in opposition to rule by a colonial power. The post-colonial states of Africa, Asia and Latin America provide many examples of this form. Colonial boundaries cut across pre-existing tribal boundaries but nations were created out of these artificial collectivities in order to rally support to the anti-imperialist cause. In this case, nationalism was an ideology created to accord with the 'artificial' boundaries of the imposed state. Nevertheless, once independence was achieved, the newly engendered national consciousness could then be used to bind the residents of the state together. However, boundaries do not conform to any meaningful 'national' divisions. This means the end product of 'bottom-up' or 'peripheral' nationalism is effectively the same as 'top-down' forms; namely, the inculcation of a national consciousness in order to bind the state together. What has happened in much of Africa, Asia and Latin America appears to have been the adoption of particular 'modular forms' derived from the European experience (Anderson 1991; Davidson 1992). In other words, post-colonial states adopted a territorial model of the nation. In the case of Nigeria, the construction of a national identity out of the structures imposed during the colonial period has proven problematic with more importance attaching to local affiliations than to a shared sense of Nigerian-ness. This reflects a polity containing within its boundaries more than 370 ethnic groups and in which the Nigerian state has failed to inculcate a strong sense of national identity among its disparate population. Conflict in Ivory Coast in 2011 also reflects an incomplete process of nation-building in a post-colonial state. Elsewhere nation-building in places such as Laos is confronted with forging a coherent nation from various layers of the past. The construction of a unified national identity among those living within the geographic space of present-day Laos (to a large extent a French colonial creation) is complicated by differing ethnic identities and by various external forces (Jerndal and Rigg 1998).

This apparent distinction between top-down and bottom-up forms of nationalism is similar to Meinecke's distinction between the cultural nation and the political nation. Writing in the early twentieth century, Meinecke suggested that the development of the former leads to the creation of the state whereas in others, the political nation (the state) precedes the inculcation of a sense of national identity (cited in Kumar 2003). Meinecke's approach is based on a distinction between *uis sanguinis* (law of blood) and *uis soli* (law of the soil) but such distinctions can be problematic giving rise to misleading ideas of 'real' nations. In the break-up of the former Yugoslavia, for some nationalists there was a tendency to portray Yugoslavia as an 'artificial' entity thereby justifying its split into its notionally 'real' constituent nations. However, Yugoslavia might be regarded as no more (or less) artificial than other national identities, including those of the nations which preceded and/or succeeded it (Djilas 2003).

The role and functions of nationalism

Because of its association with war and brutal animosity between people, nationalism is quite often seen in negative terms. Both world wars have been seen to result from nationalist territorial claims. More recently, nationalism has been seen in highly negative terms in the light of the bloody disintegration of the former Yugoslavia. With unresolved tensions in Northern Ireland, the Basque country and elsewhere, this picture of nationalism as a negative and disruptive phenomenon is reinforced. The utilization of national identity in order to justify racism and various forms of racial supremacy, whether in the era of formal colonialism or in the present-day rhetoric and actions of far-right political parties and movements, adds to the uncomfortable feelings attaching to national identity.

It is undoubtedly true that nationalism can give rise to extremely bloody consequences. The pursuit of the nationalist ideal can lead to such practices as ethnic cleansing. In the past, it served as a useful adjunct to bolster the colonization of supposedly 'inferior races'. Calls for the repatriation of 'foreigners', such as those by the British MP Enoch Powell in the 1960s and 1970s, and more recent versions of this in the rhetoric of the far right in Britain, France, Austria and other countries, represent attempts to 'purify' the territory of the nation, ridding it of 'external' contaminants. There is a danger attached to nationalism of it being either perceived as, or in reality being, an exclusivist ideology in which people who do not belong to the nation are excluded from the benefits of membership and are denied equal status. This leads to racism, ethnic tensions and, in its most virulent forms, programmes of ethnic cleansing, expulsions and repatriations. As a consequence, nationalism itself is seen as a source of evil.

In a similar vein, as Chatterjee (1993) has argued, nationalism is seen to be something of an unwanted virus spreading from the periphery and threatening to infect the core. It is seen as an ailment of eastern Europe and the periphery of western Europe (such regions as the Celtic fringe and the Basque country). It is also significant, as Seton-Watson (1977) and others have noted, that patriotism, which can be interpreted as a form of state or 'official' nationalism (Kellas 1991), is seen as 'good' while nationalism is 'bad'. Thus, defence of an existing set of national institutions can be viewed as patriotic but attempts at instilling a sense of nationhood among people with no existing political institutions tends to be seen in a negative light. Clearly, this is because events likely to lead to political instability are generally seen (at least from the perspective of those in power) as unwanted, while those which reinforce the status quo are viewed in a positive light. The sight of US citizens publicly rejoicing in New York and elsewhere following the assassination of Osama Bin Laden in Pakistan in April 2011, while chanting 'USA, USA' and waving US flags, was portrayed in various media as an understandable display of patriotism rather than as a potentially problematic manifestation of a knee-jerk triumphalist nationalism.

Related to this, Billig (1995) argues there is a strong tendency to see nationalism only in its 'hot' forms – those scenarios where it leads to violent conflict. As a consequence, nationalism is seen as a 'problem' afflicting some places but not others. While Northern Ireland and the Basque country might be portrayed as places in which nationalism is a problem, it is seldom seen as a phenomenon in relation to England or

Spain. As Blaut (1987) has pointed out, there is a tendency to focus on 'small' nationalisms rather than 'big' ones. It is perceived as a force driving secessionist movements and leading to instability. The fact that present-day states are underpinned by a sense of nationhood, subtly instilled through various mechanisms, is ignored. This tendency to see nationalism as a phenomenon of the periphery is bound up with imperialism and conquest: 'those whose identities are rarely questioned and who have never known exile or subjugation of land and culture, have little need to trace their 'roots' in order to establish a unique and recognizable identity' (Smith 1986: 2). Within the United Kingdom, nationalism tends to be seen as something which is peculiar to the Scots, Welsh and Irish while curiously not afflicting the English. Indeed, as Eric Hobsbawm (1990) has noted, the very term 'English nationalism' sounds odd and tends to be associated with the far right of the political spectrum, though ideas of national chauvinism and jingoistic behaviour continue to exist among more mainstream elements.

As Anthony D. Smith (1986) points out, most historians appear to view nationalism in negative terms. Eric Hobsbawm is a case in point. To him and others, nationalism appears to be an intrinsically 'bad thing'. In part, it might be argued, this reflects the quest for academic objectivity and detachment. It might also be argued that it reflects certain academics' disdain for what is seen as vulgar, sentimental and, perhaps, irrational behaviour. In addition, nationalism is often (though by no means always) associated with the political right and, as already suggested, can carry with it exclusionary ideas acting as an excuse for racist and xenophobic behaviour. While Hobsbawm's laments over the bloody effects of nationalism appear to reflect a commendable vision of an egalitarian world shorn of national chauvinism (and shaped by his own multinational background), they also play down the understandable appeal of nationalist rhetoric to people whose rights and freedoms have been trampled on. Related to this is the presumption that extreme nationalism is aberrant or irrational behaviour (Kellas 1991). This is akin to the view that nationalism is essentially a disease, the implication being that it needs to be eradicated (Miller 1997). It might be argued that extreme nationalism (however that might be defined) may reflect a perfectly rational set of motives – defence of home, job, security, power, etc. Those who have pursued Irish unity through military means may well perceive (rightly or wrongly) continued British rule in the north of Ireland as being against their better interest. Thus, there may well be more than a supposedly irrational attachment to territory or a misty-eyed romanticism underlying nationalist movements (though these are undoubtedly quite often present and indeed are often invoked as part of the 'nationalizing' strategy). For

Tom Nairn (1997), a world of an increasing number of nations is not 'unreal' or aberrant. Rather, it points to the powerful meanings which attach to national identity, meanings which arise for a variety of reasons. While there are many negative phenomena associated with nationalism (such as racism and xenophobia), Nairn asserts that the nation is far from dead and it may offer hope to some, thus suggesting that a total demonizing of the phenomenon is neither apt nor useful.

It is possible to see national differences in a more positive light. A distinction can be made between an exclusive or narrow conception of nationality and a more inclusive or broader perspective. The former revolves around racism and xenophobia while the latter is a pluralistic version. This can be related to another distinction: that between a nationalism which is chauvinistic and another which asserts cultural distinctiveness but which eschews assertions of national superiority. The acceptance of national differences in terms of such aspects as language and other elements of culture can be seen to contribute to international diversity and cultural enrichment. This contrasts with national chauvinism which asserts that *one* nationalism is superior to all others. Manent (1997) argues that the nation can be the basis for a set of positive values and a framework for meaningful democratic institutions. Rather than being automatically associated with regression, nationalism might be linked to progressive democratization.

Leaving aside the more overt manifestations of national identity, the phenomenon is reinforced in ways which seem quite ordinary and mundane and appear, on the surface at least, to have nothing to do with politics. Many people feel 'English', 'Irish', 'Spanish' and also feel they are different in many respects from members of other nations. This does not necessarily lead to conflict but merely to the recognition of differences. Conflicts only arise in situations where territorial expansion into the 'space' of others occurs or where national chauvinism exists where members of a national group perceive themselves to be superior in some way to members of one or more other national groups. None of this is to ignore the fact that people may identify with their locality, county, region, or with more than one nation, as suggested earlier. There are levels of territorial identity and the intensity of feeling will vary depending on the specific context. Regardless of the academic 'opinion' of it, nationalism is a phenomenon requiring serious study and 'merely to inveigh against nationalism does little to help the human race' (Seton-Watson 1977: xii). It might be appropriate to view nationalism itself as a neutral phenomenon, one which can be put to use for what might be seen as either progressive or conservative ends.

Irrespective of the view that is held on the origin of nations, and regardless of whether it is seen as a progressive or regressive phenomenon, nationalism serves a number of purposes in modern societies. The nation can serve as a convenient mobilizing tool to help the state to achieve its objectives. In this way, it is useful in maintaining hegemony and legitimacy for the state. An alternative set of functions is more subversive in that nationalism can be used as a mechanism for righting other wrongs. Thus, greater democracy or equality for marginalized groups may be fought for under the banner of nationalism. It is clear that nationalism is a very useful mobilizing force, particularly if it can be shown that 'the nation' is under threat. It has already been suggested that nationalism as an ideology has served to perpetuate certain forms of territorial control. In the 1990s and again in the early 2000s Saddam Hussein utilized a version of nationalist rhetoric in urging the people of Iraq to support him in his defence of the nation against the external threat posed by the United States. Similarly, US regimes can refer to a defence of American interests as a mechanism for mobilizing public opinion over acts of aggression towards other countries though, in reality, what may be happening is a defence of broader US strategic and economic interests.

National identity can be used to maintain political hegemony. A political elite can invoke nationalist rhetoric in order to maintain its own hegemonic control within its territory. Nations can be constructed around particular sets of ideas. Thus, the Nigerian writer Wole Soyinka describes a nation as 'a gambling space for the opportunism and adventurism of power' (1996: 286). He argues that the geographic spaces that comprise present-day African states are spaces in which people are marginalized and exploited, while those doing the exploiting invoke the rhetoric of nationalism to mask their otherwise naked display of power. The distinctions between government, state and nation are sometimes deliberately blurred by governments as a mechanism for undermining opposition. Governments may occasionally claim that opposition to them is unpatriotic. In this way, opposition to US foreign policy is sometimes portrayed as un-American. To oppose its foreign policy (a criticism of government) is sometimes used by the administration as an indication of a lack of patriotism (support for the country) in an attempt to undermine the validity of the arguments being used. However, the majority of such criticism, whether from the left or the right, is not about undermining the nation, but rather about aspects of state policy or the nature of the state. It is not necessarily opposition to the United States as a political entity (although it might be). Indeed, right-wing opposition within the USA is often of a 'patriotic' nature arguing against state intrusion but supporting 'American values'. To be labelled as anti-American is a useful method of minimizing protest against unpopular policies. If opposition to specific measures or a specific regime can be presented as opposition to the nation, the arguments may appear less popular. The nation, as a concept, is thus a very useful means of political control. It enables those in power to get people to identify with the nation and specific measures can be justified on the nation's account. Thus, people are encouraged and expected to 'rally round the flag' as exemplified in the famous words of former US president John F. Kennedy: 'ask not what your country can do for you, ask what you can do for your country'. In the aftermath of the 9/11 attacks on the United States, then president George W. Bush declared that people (in the United States and elsewhere) and other countries were 'either with us or with the terrorists', thereby foreclosing any more meaningful debate over the issue and reducing it to a simplistic 'us' versus 'them' scenario.

Nationalism can be sold as a sop to those who feel alienated. In this way, the poor and the unemployed can take pride in their nationality. People are urged to look on 'higher' things rather than the mundane events of everyday life. The 2011 royal wedding of Prince William (second in line to the British throne) in the UK was heralded as a day of great national celebration and pride serving, perhaps, to take citizens minds off more serious and pressing problems of job losses and foreign military interventions (Box 4.4). People may be elevated above the tedium and hardships of the present, thereby rendering them more subservient to those who rule in the name of the nation. The nation is, thus, an agent of legitimation and mobilization, a means by which powerful elites can retain or gain power. Working-class Protestants in Northern Ireland can feel proud of their religion and their British heritage. In this way, they can be mobilized as a political force in defence of what they see as their right to remain within the UK. As Soyinka (1996) argues, clearly with his native Nigeria very much in mind, political elites can invoke nationalist rhetoric and play on what might otherwise be regarded as relatively meaningless differences in a strategy of divide and rule. As a result, many African countries (and others elsewhere) retain significant internal divides, thereby impeding action which might otherwise lead to some form of more progressive politics. The divide and rule strategy can be used in cases to create overt strife which in turn can be used to justify harsh measures to restore stability. Hegemony can sometimes, perversely, be retained through the invention of instability. As well as political elites, particular sectional interests may be advanced through an invocation of the national so that, for example, fox-hunting in Britain is held up by campaign groups as an intrinsic element of Britishness linked to the countryside, the land, identity and everyday life (Wallwork and Dixon 2004).

Box 4.4 Banal nationalism: Britain's royal wedding

Prince William is second in line to the British throne and his wedding to Kate Middleton in April 2011 was presented as a joyous occasion allowing the nation to come together and celebrate this event. A national holiday was declared for the day of the wedding, many public buildings, shops and houses were bedecked in flags and bunting, and street parties were held up and down the country to celebrate the event (Figure 4.8). It was used to generate a feel good factor, promoting a positive affirmation of national identity. The wedding was also, of course, the catalyst for a range of merchandizing and marketing opportunities, as well as generating tourist revenue and projecting a positive image abroad. In doing this, it could be interpreted as an event serving to deflect attention away from serious socio-economic difficulties, to rally support (or at least deflect criticism) for an incumbent government.

However, the wedding celebrations were not unequivocally successful in producing a unified and celebratory response. Threats to disrupt the event were alleged to have been made by radical Islamist groupings, by Irish republicans and by anarchist and anti-royalist factions. However, those few dissident voices which were heard in the mainstream media tended to be treated as marginal, rather than their arguments being given serious consideration. Much media coverage from supposedly neutral journalists was sycophantic and reflected a knee-jerk nationalism. It is also ironic that the multi-ethnic nature of Britain's royal family (closely related to other European monarchical families) tends to be ignored and instead the monarchy is presented as a quintessentially British institution.

Figure 4.8 Flags celebrating Britain's royal wedding, 2011 (Source: Author)

In a different way, nationalism can be seen as a useful tool in attempting to redress what are seen as social injustices. In many instances, what appear to be disputes centring on national or ethnic tensions may be more to do with the marginalization of particular groups and may be connected with social exclusion as much as with nationalism per se. Tamil separatism in Sri Lanka may owe its strength more to the marginalization of Tamils as with any innate sense of wishing for full political autonomy so that the latter may, in effect, be a device for redressing the former. Kurdish calls for an independent Kurdistan reflect not just a territorial wish to have their own political space, but also the social, economic, political and cultural marginalization of Kurdish people in the countries in which they find themselves, most notably Turkey and Iraq. Viewed in this way, nationalism can be seen as a territorial strategy useful in achieving social and economic objectives.

In more mundane ways, certain events can be used as a means of engendering a feel good factor in relation to the nation. The rescue of 33 trapped coal miners in Chile in November 2010 (beamed across the world live on television) was projected through the frame of Chilean nationalism with national flags flown at the rescue site and by the miners underground. The singing of the national anthem by the miners added further to this conversion of a heroic event into a form of political theatre, engendering positive feelings and forging an upbeat national narrative. The Chilean president, Sebastián Piñera, was quick to capitalize on this, being present to greet each miner as they reached the surface following their lengthy confinement. The positive virtues associated with the rescue were useful political capital, certainly more so than a questioning of the mining industry regulations and procedures may have contributed to the men's entrapment in the first place.

Finally, as the above examples may suggest, it might be argued that nationalism is in some ways a necessary precondition for democracy. If there is a sense that people should have a say in how they are ruled, this implies some idea of who the 'people' actually are. Despite the view commonly held that nationalism and democracy are mutually antagonistic, writers such as Nodia argue that some sense of nationalism, based on a collectivity of people within a defined territory, underpins all forms or attempts at democracy. While there is nothing natural or inevitable about nations, it is the case that 'nationalism is the historical force that has provided the political units for democratic government' (Nodia 1994: 7).

Summary

This chapter has provided an introduction to the concepts of the nation, national identity and the associated territorial ideology of nationalism. Theories on the origins of nations and the role and purpose of nationalism have been reviewed. Regardless of the debates surrounding the origins of nations and the variety of views surrounding the phenomenon of nationalism, it is clear that nations do exist. People see themselves as belonging to a nation. Whether this is viewed as rational or irrational, positive or negative, progressive or regressive, it remains an indisputable phenomenon. It is equally clear that nations are 'produced' rather than being 'natural'. It follows that a national identity is a constructed one and, consequently, nationalism serves particular functions at specific times. Despite the recognition of nationalism as a historically contingent phenomenon which serves particular ends, 'none of this makes the nation "unreal" for an ordinary man (sic) born into a concrete society, culture, and state, and faced with concrete choices on the social and political as well as the spiritual and existential planes. The nation need not be "rational" in order to be "real"' (Nodia 1994: 11). Equally indisputable are the powerful connections between national identity and territory. These links between place and nation are explored more fully in the next chapter.

Note

1   Lyrics available at www.lyricstime.com

Further reading

There is a wealth of literature dealing with nations and nationalism. Some particularly important contributions from key theorists are listed below, together with helpful overviews of key debates and some useful edited collections.

Anderson, B. (1991) *Imagined Communities. Reflections on the Origin and Spread of Nationalism*, London: Verso.

Billig, M. (1995) *Banal Nationalism*, London: Sage.

Chatterjee, P. (1993) *The Nation and its Fragments. Colonial and Post-Colonial Histories*, Cambridge: Cambridge University Press.

Davidson, B. (1992) *The Black Man's Burden. Africa and The Curse of the Nation-State*, Oxford: James Currey.

Delanty, G. and Kumar K. (eds) (2006) *The Sage Handbook of Nations and Nationalism*, London: Sage.

Gellner, E. (1983) *Nations and Nationalism*, Oxford: Blackwell.

Gellner, E. (1994) *Encounters with Nationalism*, Oxford: Blackwell.

Gellner, E. (1997) *Nationalism*, London: Wiedenfeld and Nicholson.

Guibernau, M. (2007) *The Identity of Nations*, Cambridge: Polity Press.

Harris, E. (2009) *Nationalism. Theories and Cases*, Edinburgh: Edinburgh University Press.

Hobsbawm, E. (1990) *Nations and Nationalism since 1780. Programme, Myth, Reality*, Cambridge: Cambridge University Press.

Kumar, K. (2003) *The Making of English National Identity*, Cambridge: Cambridge University Press.

MacLaughlin, J. (2001) *Reimagining the Nation-State. The Contested Terrains of Nation-Building*, London: Pluto Press.

Miller, D. (1995) *On Nationality*, Oxford: Clarendon Press.

Nairn, T. (1997) *Faces of Nationalism. Janus Revisited*, London: Verso.

Seton-Watson, H. (1977) *Nations and States. An Enquiry into the Origins of States and the Politics of Nationalism*, London: Methuen.

Smith, A. D. (1986) *The Ethnic Origins of Nations*, Oxford: Blackwell.

Smith, A. D. (1991) *National Identity*, London: Penguin.

Smith, A. D. (1998) *Nationalism and Modernism. A Critical Survey of Recent Theories of Nations and Nationalism*, London: Routledge.

Smith, A. D. (2001) *Nationalism: Theory, Ideology, History*, Cambridge: Polity.

5 Nationalism, territory and place

While not everyone subscribes to the idea of the nation in the same way or to the same extent, nevertheless, for many people their national identity is a significant component of their make-up, even if this is only apparent in seemingly mundane ways. The sight of football players and fans proudly singing their national anthem while waving or saluting 'their' flag, perhaps with tears in their eyes, is a powerful reminder of the pervasiveness of national identity. This reflects the success of efforts to create a strong national consciousness. This sense of the nation is constantly constructed and re-constructed in ways sometimes overt and sometimes banal, as suggested in the previous chapter. In order to do so, nations require features which can be utilized in the process of affirming or re-affirming nationhood. While the importance of history is clear, what is less widely acknowledged is the significance of geography. There is, quite clearly, a need for a national past which is seen to provide the glue that holds the nation together. In tandem with this national past is a national geography built around particular places and utilizing explicit territorial allusions.

While people and events are put into the service of nation-building and affirmation, place also assumes significance. Places are, by their nature, geographic entities: people live in places and events occur in, or are associated with, places. In this way, a 'national' history works in conjunction with a 'national' geography to present a vision of the nation. Particular sites or particular landscapes – spaces and places – become imbued with meaning:

> a certain tradition of images, cults, customs, rites and artefacts, as well as certain events, heroes, landscapes and values, come to form a distinctive repository of ethnic culture, to be drawn upon selectively by successive generations of the community.
>
> (Smith 1991: 38)

Within this mix, diverse elements such as battles, language and landscape come together to produce the nation and are utilized in order to preserve its territorial integrity. In this regard, place can be invoked in two key ways: first, in terms of generic landscapes and, second, in the significance seen to attach to specific places. This chapter explores the use of places, as well as people and events, in the making and re-making of the nation. First, the importance of a 'national' history is assessed. Subsequently, the connections between particular places and the nation are explored. The chapter continues by considering the significance of territory in the construction of the nation in four rather different contexts: the Balkans, Ireland, Israel/Palestine and England/Britain. The key focus in each of these examples is the importance of territorial imagery in the construction and reproduction of the nation. In situations of overt conflict, territorial strategies are utilized in order to reinforce or to resist a particular political configuration. Territory, it is argued, is important both symbolically and in practical terms.

The importance of history

All nations require a past to justify their current existence and to provide a rationale for territorial claims. A version of the nation's past needs to be brought into being. 'National' histories tend to present a relatively seamless narrative through which the members of the nation can trace their collective past. This is not the same as saying that an 'accurate' version of history is important. As Hobsbawm suggests a 'suitable' past is required and 'if there is no suitable past, it can always be invented' (1998: 6). Fact, folklore and fiction combine to produce and reproduce a sense of nationhood and myths and legends are an important part of nation-building. Indeed, 'inaccurate' histories are perhaps crucial and the blurring of myth and reality is central to the nationalist imagination: 'myths are building blocks of nationhood' (Perica 2005: 130). As Ernest Renan observed towards the end of the nineteenth century, 'getting its history wrong is part of being a nation' (1882, quoted in Hobsbawm 1990: 12). In fact, it may be *necessary* to get it wrong. From a functionalist perspective, these elements of the past, whether real or invented, are utilized for contemporary purposes.

> A claim to national independence does not fall simply because its legitimising version of national history is partly or wholly untrue – as it often is. The sense of belonging to a distinct cultural tradition . . . can be subjectively real to the point at which it becomes an objective socio-political fact, no matter what fibs are used for its decoration.
>
> (Ascherson 1996: 274)

Once particular political territorial configurations take root and systems, myths and practices become accepted, then the nation assumes a mantle 'if not of immemoriality, at least of almost inevitable and organic development' (Davies 2000: 54). As noted earlier, particular people may be invoked in the propagation of the nationalist myth. Certain key historical figures are assumed to embody the nation such as Nelson in England, Lincoln in the USA, William Wallace in Scotland, Owyn Glyndwr in Wales. These figures are mythologized, or practically invented in some instances, and become symbols of the nation's past when they fought for its freedom or performed glorious deeds on its behalf. It is not just obvious political figures who can be taken as symbolizing the nation. Even contemporary personalities from beyond the world of politics or the military can be invoked. Sportspeople are often seen as carrying the mantle of the nation and their actions may be interpreted as symbolizing the national spirit. Indian cricketers could be said to carry the hopes of the nation on their shoulders, particularly when playing Pakistan, while Spain's victory in the 2010 football World Cup brought huge crowds onto the streets of Madrid and other Spanish cities to celebrate 'their' win.

Just as certain individuals are taken as emblematic of the nation, particular events may be seen as vital to the national project. Scottish people are assumed to identify with the Battle of Culloden while the Highland Clearances are seen as an event which exemplifies the suppression of Scotland by England. The victory of William of Orange (a Dutchman!) over King James at the Battle of the Boyne in 1690 (well before the partitioning of the island) is seen as a defining moment in Irish history underpinning the right of the unionist people of Northern Ireland to remain under British rule. Sites of major battles, as well as buildings associated with key individuals, become national monuments – places seen as being of vital historical importance and, hence, of crucial significance to the nation's present (Figure 5.1). In a world ever more fully in thrall to heritage in its various guises, the preservation or construction of a national heritage built around people and events is of ever-increasing significance (Lowenthal 1998; Graham *et al.* 2000).

It follows from the above that the veracity of the role of individuals or the nature of key events is not what is important. Rather, it is the mythological interpretation which is placed upon them. Events and people become 'traditionalized' in order to celebrate the nation. In this way, it can be said that nations are constructed through the invention of tradition whereby particular customs or events are portrayed as stretching back into antiquity to a time immemorial, to the primordial nation (Hobsbawm and Ranger 1992). Examples of this invention include the Scottish kilt, Welsh Eisteddfodd and the British coronation ceremony, all of which

Figure 5.1 Memorial to Irish rebellions against British rule, Skibbereen, County
Cork, Ireland (Source: Author)

are nineteenth-century inventions (although they may be partly based
on earlier formats). All of this implies a construction of the nation out
of various historical and geographical fragments. In the flowery prose
of Anthony D. Smith, nationalism provides 'cognitive maps and historical
moralities for present generations, drawn from the poetic spaces and
golden ages of the communal past' (1991: 69). Once again, this highlights

the emphasis on the invocation of both the geographic and the historical in the reproduction of the nation.

Territory and nation

Within the discourse of nationhood, it is obvious that territory is of huge importance; indeed, 'only territory provides tangible evidence of the nation's existence' (Herb 1999: 10). However, it is not just the land itself that is important; historical fact and myth concerning particular places is a key element in the national imagination. History and culture, as Etherington (2010) suggests, mediate between the nation and its territory: the territory belongs to the nation and the nation belongs to the territory. The importance of key individuals and events in nation-building has already been noted, with such individuals and events usually being associated with particular places. In addition to these indirect place-associations, there are two main ways in which territory features in nationalist narratives. First, there are numerous references to the 'generic' territory of the nation. Allusions to the national soil abound within such discourses. The second territorial element is the importance attaching to specified places. For example, past Serbian occupation of present-day Kosovo serves as a useful justification for territorial control over an area which is almost exclusively occupied by ethnic Albanians. The connections between people and place, discussed in Chapter 2, suggest that a sense of place, when linked to the political project of nationalism, results in the sustaining of evocative images of place which come to symbolize the nation. Of course, the nature of nationalism is such that the same 'facts' may be interpreted radically differently by different people. National heroes may be villains in the eyes of others, while particular landscapes may be a source of pride for some but shame for others. This dissonance means it is highly unlikely that there will be universal agreement over what constitutes an authentic interpretation of the national past – that which is presented as historically accurate and authentic will be seen by some as distorted and misleading. The idea of an essentialized national heritage built on the bedrock of an ancient and immutable cultural (and territorial) core is an unattainable (and probably undesirable) goal. Nevertheless, for some there is the need to hang onto this vision in which landscapes and places are central and which serves as a bulwark against modernity and other threats to the integrity of the nation. In all of this we can see what Williams and Smith (1983) have referred to as the national construction of social space.

Certain places may acquire huge symbolic importance. Particular parts of the national territory may acquire a significance as the presumed 'zone of origin' of the nation: its original heartland. Thus, the Canadian north with its vast, remote, rocky and forbidding image is seen as symbolizing an independent Canada (Figure 5.2). The 'taming' of the American west means that not just the 'pioneers' heading westwards, but also the landscapes through which they travelled assumed significance in the nation-building project. In versions of Welsh nationalist discourse, the mountains are seen as the heart of the nation, somehow symbolizing a Wales untainted by outside, specifically English, influences. The Welsh nationalist party, Plaid Cymru, chose as its original symbol an idealized representation of mountains. A leading figure in the party during its early years suggested the Welsh mountains are 'the perpetual witnesses of our history, and the unchanging background of our language' while another nationalist described them as 'the bread of life, and . . . a holy sacrament in which lives are woven into its essence' (cited in Gruffud, 1995: 224, 228). In this way, 'remote areas away from the anglicizing influences of the accessible lowlands harboured the "national character" and nationalist politics, therefore, centred on the defence of that cultural integrity' (Gruffud 1995: 236). Indeed, in some discourses, the land itself assumes the role of a sentient being whose 'memories are sacred, its rivers are full of memories, its lakes recall distant oaths and battles' (Williams and Smith 1983: 509). Moreover the land acts as a defensive bulwark so that, for example, in some discourses Basque cultural purity is deemed to be safeguarded by its mountainous terrain (Lowenthal 1994). These discourses can be seen as attempts to bind the nation and its people to the physical characteristics of the land (Etherington 2010). Political leaders are also wont to invoke the land in attempts to garner support. Former US president Ronald Reagan's sense of American territorial inevitability provides a classic example:

> I have always believed that this land was placed here between the two great oceans by some divine plan. It was placed here to be found by a special kind of people – people who had a special love of freedom and who had the courage to uproot themselves and leave hearth and homeland and come to what in the beginning was the most under-developed wilderness possible.
>
> (Ronald Reagan in 1980, quoted in Cresswell 2004: 72)

While these allusions to generic landscapes and features are important, specific places also acquire significance. In this way, the White Cliffs of Dover are taken as one of the symbols of England and, by extension, Britain: the embodiment of the nation in a place. In Eritrea, the currency is named after a place – Nakfa – which was heavily attacked by Ethiopian

Figure 5.2 Rocky Mountains, Canada (Source: Author)

forces during the struggle for independence in the 1980s. The Basque town of Gernika/Guernica has a highly significant role in the Basque conflict, particularly as the town was bombed by the German Luftwaffe in 1937 at the behest of Spain's fascist dictator Franco. The town itself, and the symbolism attaching to it, render it a pre-eminent place in Basque identity (Raento and Watson 2000). Elsewhere, ideas of heroic defeats and territory lost or unrecovered imply unfinished business and can be used to remind people that the nation is incomplete and will remain so until the 'lost' territory is returned to the fold.

The importance of territory and of specific places is emphasized in many national anthems. These pieces of music, the most overt means by which the nation is symbolized, quite often contain territorial references. These may take the form of generic allusions to soil and land or more specific references to particular places or landscape features such as mountains or rivers. In this way, songs seen as the nation's musical signature often have a strong territorial base, evoking images seen as part of the essence of the nation. Hobsbawm has written of the attempt to inculcate a sense of Austrian-ness through the use of territorial imagery in the first short-lived national anthem of post-World War I Austria (a remnant of the former Habsburg empire). In Hobsbawm's words, this anthem involved 'a travelogue or geography lesson following the alpine streams down from glaciers to the Danube valley and Vienna' (1990: 92).

The importance of territory and of specific places continues to be emphasized in many current national anthems. Austria's current anthem provides an example of the use of generic landscape features as symbols of the country with its references to a 'land of mountains, land of rivers, land of tillage, land of churches'. In a similar vein, 'Flower of Scotland' evokes a rugged rural landscape of hills and glens, one which is defended against the marauding external foe (England). These examples also demonstrate the elevation of quite ordinary landscapes into something almost sacred. There is nothing intrinsically unique about mountain ranges, coastlines or woods, but in the nationalist imagination these features assume emblematic status. Norway's anthem refers to its citizens' love of this country, a 'land that rises rugged, storm-scarred, o'er the ocean with her thousand homes'. For sheer elevation of territory to a mythical status, Chile's anthem takes some beating:

> Chile, your sky is pure blue
> So sweet as the breezes that roam
> Over your fields, embroidered with flowers
> That angels might make thee their home

Such hyperbole serves to manufacture an almost magical landscape. Similar sentiments abound in Denmark 'whose charming woods of beeches grow near the Baltic strand'.

While the above refer to landscape features in quite general terms, many anthems refer to specific places seen as integral to, or in some way the embodiment of, the nation. Anthems may sometimes make what are, in effect, specific territorial claims:

> Proudly rise the Balkan peaks
> At their feet the blue Danube flows
> Over Thrace the sun is shining
> Pirin looms in purple glow

In this verse, Bulgaria lays claim to its defining territory with mountain and river signifying the nation. Of particular significance here is the reference to Thrace, an ancient territory now largely in Greece. In this way, Bulgaria lays claim, emotionally at least, to this area now lying within the borders of another country. Similarly, the anthem of Croatia refers to rivers:

Sava, Drava, keep on flowing,
Danube, do not lose your vigour,
Deep blue sea, go tell the whole world,
That a Croat loves his homeland

The Drava, Danube and Sava form part of the country's borders with Hungary, Serbia and Bosnia-Herzegovina respectively. Throughout many national songs, the defence of the homeland is a recurring theme: the salvation, or maintaining, of territorial integrity is seen as crucial rather than the attaining of democracy or a particular form of government. In this way, the Mexican anthem addresses the territory directly and assures it that 'should a foreign enemy dare to profane your soil with his tread know, beloved fatherland, that heaven gave you a soldier in each of your sons'.

These examples also demonstrate the elevation of quite ordinary landscapes into something extraordinary. Beyond the celebratory form of the anthem, writers and politicians are wont to make explicit territorial allusions when referring to their country. The early twentieth-century travel writer H.V. Morton encapsulates this notion of the land, the rural and farming as integral to the idea of the 'nation' (in this case, England):

> There rose up in my mind the picture of a village street at dusk with a smell of wood smoke lying in the still air and, here and there, little red blinds shining in the dusk underneath the thatch . . . The village and the English countryside are the germs of all we are and all we have become: our manufacturing cities belong to the last century and a half: our villages stand with their roots in the Heptarchy.
>
> (Morton 2000: 1–2)

Not surprisingly, politicians are wont to use such sentiments in public pronouncements with overtly rural images used to promote a sense of national identity. In this way, in 1937 former British prime minister Stanley Baldwin spoke of:

> The sounds of England, the tinkle of the hammer on the anvil in the country smithy, the corncrake on a dewy morning, the sound of the scythe against the whetstone, and the sight of a plough team coming home over the brow of a hill, the sight that has been seen in England since England was a land, and may be seen in England long after the Empire has perished and every works in England has ceased to function, for centuries the one eternal sight of England.
>
> (cited in Paxman 1998: 143)

More recently, in 1993 then prime minister John Major (echoing the writer George Orwell many years previously) invoked a vision of Britain with 'old maids cycling to holy communion through the morning mist' (cited in Paxman 1998: 142).

While much of these claims may be seen as reflecting the ideas of cultural nationalists and the intelligentsia, ideas then put into the service of political leaders, a more proletarian territorial rhetoric may also be found. The American folk singer Woody Guthrie used an explicitly territorial sense of ownership to demonstrate that ordinary working-class people should have a stake in the nation – that the land could be recaptured:

> This land is your land, this land is my land
> From California, to the New York Island
> From the redwood forest, to the gulf stream waters
> This land was made for you and me[1]

In highlighting the importance of particular places, it is illuminating to present examples of the ways in which geography and a sense of place become integral to the creation and sustainment of a national imagination. The four examples which follow are derived from somewhat different political contexts. The first three relate to regions in which territorial issues remain unresolved. In the Balkans, specific territories have assumed symbolic importance for different ethno-national groupings. Irish identity is constructed in the light of a colonial struggle against Britain. The significance of territory in the case of Israel/Palestine lies in two national groups laying claim to the same national space. The final example pertains to a more banal form of national identity with a consideration of territorial imagery in the construction of English identity – a phenomenon complicated by confusion between English and British.

Territory, place and nationalism in the Balkans

The state of Yugoslavia was created at the end of World War I out of territory which had long been contested between the Ottoman and Austro-Hungarian empires. It is a region containing a complex variety of ethnic groups which have a history of both antagonism and peaceful inter-mingling and inter-marriage. In the 1990s, Yugoslavia began to fall apart with its constituent 'nations' endeavouring to go their own ways. The result was the eventual creation and recognition of Slovenia, Croatia, Bosnia-Herzegovina and Macedonia as independent states with Yugoslavia reduced to Serbia and Montenegro. Subsequently, Montenegro split from Serbia in 2006, and in 2008 Kosovo also seceded from Serbia (Figure 5.3). The nature of Balkan history means that territorial divisions between distinct groupings are not easy to identify, with many competing claims over particular places (Storey 2002).

The disintegration of Yugoslavia was far from peaceful and was highly contested. Violent conflicts erupted in Croatia, Bosnia-Herzegovina and, later, Kosovo. These conflicts were characterized by attempts by various groups to eradicate other ethnic groups from 'their' territory. This strategy of ethnic cleansing was built on an essentialist version of defining ethnonational identity and, quite literally, clearing the territory of those possessing a supposedly 'different' identity. Armed movements claiming to be representing Serbs, Croatians and Bosnians (predominantly Muslims) tried to carve out spaces which they could call their own in the ruins of the former federal state. Viewed from the outside, it is easy to argue that such a strategy is both dangerous and simplistic. However, for those directly involved, the control of territory was seen as an essential element in the conflict. These wars resulted in hundreds of thousands of deaths, ruined towns and villages, and infrastructural breakdown with the struggle over Bosnian territory being particularly violent (Figure 5.4) (Toal and Dahlman 2011). Despite the area's complex multicultural

Figure 5.3 The Balkans

Figure 5.4 War-damaged buildings in Mostar, Bosnia-Herzegovina (Source: Author)

history, reductionist interpretations of identity led to attempts to assert territorial control through the elimination of 'others'. Nationalist rhetoric, and the associated desire to control particular portions of territory in the name of a specific group, hardened divisions which were of relatively minor significance only a few years previously when Yugoslavia was a federal (and communist) republic under one-party rule. Once the conflict was underway however, it became impossible to reverse, as it drew upon and reinforced group identities and competing territorial claims, underpinned through reference to historical myths and forms of boundarymaking. In order to achieve peace, areas were mapped and information gathered on the various ethno-national groupings living in different localities. In this way, territory was designated 'Serb', 'Muslim', 'Croat' and so on, with lines dividing towns and cities into different zones. After a number of failed alternatives, the Dayton Agreement of 1995 divided Bosnia-Herzegovina into two autonomous units: a Muslim-Croat Federation and a Bosnian Serb Republic (*Republika Srpska*) (Figure 5.5). While this solution was lauded in many circles, critics have argued that it is inherently unstable and that it also reinforces ethnic divisions rather than rising above them (Campbell 1999; Robinson and Pobric 2006).

While conflict raged over enclaves of Serbs and Croatians within Bosnia and the designation of particular zones as 'belonging' to one or another ethnic grouping, there is another vitally important territorial

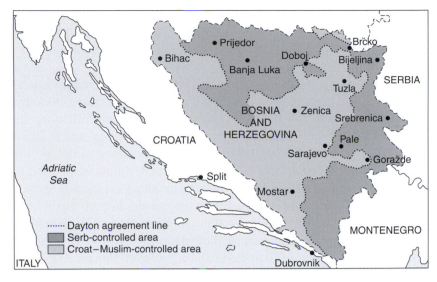

Figure 5.5 Bosnia-Herzegovina

component to the conflict which extends beyond the present-day ethnic composition of the population. Central to these disputes have been claims and counter-claims to places seen as historically belonging to one group or another. Underlying this ethnic tension, Serb 'geography' is of prime importance. White (1996) has highlighted the significance to Serb nationalists of places which are not actually in present-day Serbia. Thus, Serb folk songs make mention of places in neighbouring states of the former Yugoslavia and beyond – Croatia, Bosnia-Herzegovina and even in Greece and Albania. Songs refer to battle sites, to mountains and to rivers in territory outside Serbia's current borders. White quotes Stevan Kacanski's 1885 verses:

> Round the Struma and the Vardar
> a lovely flower blooms,
> it is the flower of the Serbian tsar:
> the holy blossom of Tsar Dusan.
> Round the Struma and the Vardar
> bloom the flowers of the Serbian tsar

In this extract, two rivers are used to evoke an image of a Serb heartland, but the Struma and Vardar run through the territory of present-day Macedonia. Although the sub-dividing of the former Yugoslavia along broad ethnic lines might be seen to reflect current ethno-national distributions of population, it ignores the fact that certain places are imbued with meaning for Serbs, even though few may actually live in these places. A concentration on the ethnic composition of a particular locality fails to take account of the symbolic meaning of particular places (White 1996, 2000).

Of particular significance in this regard is Kosovo which, within federal Yugoslavia, had been a region in Serbia but with considerable autonomy. Serbian attempts to roll back its relative independence following the collapse of the federal Yugoslavia were met with political and military opposition, with the Kosovo Liberation Army claiming to defend the ethnic Albanians against the colonizing Serbs. Ultimately, this led to outside intervention with NATO launching air attacks on Serbia and the deployment of a peace-keeping force on the ground. In Serb mythology, their defeat in a battle by the Turks at the Battle of Kosovo Polje in 1389 is seen as a formative element in their national identity. To concede control of the location of their heroic struggle, even though few Serbs live there, would be seen as sacrificing an integral part of Serb territory, and through it, Serb identity. It would be construed as a betrayal of their own history. The symbolic significance attached to Kosovo proved a useful device through which the then Serbian leader Slobodan Milosevic tried to sustain political support built around nationalist–territorial rhetoric. In a speech at the battle site of Kosovo Polje in 1987, Milosevic reminded the region's Serbs:

> This is your land. These are your houses. Your meadows and gardens. Your memories. You shouldn't abandon your land just because it's difficult to live . . . It has never been part of the Serbian character to give up in the face of obstacles, to demobilize when it's time to fight . . . You should stay here for the sake of your ancestors and descendants.
>
> (Slobodan Milosevic, cited in Harris 2009: 132–133)

Similarly, Macedonia was the centre of a medieval Serbian empire but in some discourses it is also deemed historically to be an integral part of both Greece and Bulgaria, while the country also has a sizeable ethnic Albanian minority. In fact, Greece objected to the use of the name Macedonia as the title of the newly independent state. Hence, its official designation (in English) is the Former Yugoslav Republic of Macedonia, as noted earlier. The importance of territory as a site of key events in nation-building is augmented by associations with national heroes. Alexander the Great, a Greek hero, came from what is present-day Macedonia, as did the Serb hero Prince Marko so 'to say that Macedonia is not rightfully part of Serbia is to say that Prince Marko was not really

Serbian' (White 1996: 1). Further confusing the issue here is that the Bulgarians also claim Prince Marko as a national hero. In such a scenario, the construction of separate identities becomes hugely important. For the purposes of Macedonian nationalism, it was necessary to accentuate distinctions between Macedonians and others, particularly Bulgarians and Greeks, emphasizing a separate language among other things. Deep and distinct histories are also required and claims are made about the history and continuity of a region known as Macedonia, home to a people seen to be ethnically distinct and deserving of their own state (Brunnbauer 2005).

In summary, the ethnic conflict in the former Yugoslavia is a clear example of the symbolic importance of territory and it demonstrates the manner in which territorial arguments can be utilized in order to gain, or to maintain, political dominance. While some discourses reduce the Balkans and other conflicts to deep and historical ethnic hatreds, these simplistic assertions mask the material concerns of people while allowing political leaders to gain credibility and stature through the stoking of those myths (Kolstø 2005; Gagnon 2006; Jeffrey 2006).

Four green fields

Territory and territorial imagery have played a prominent role in the creation of Irish identity while territorial strategies play a part in the ongoing contestation over the north of Ireland. Traditionally, Ireland has portrayed itself (and been portrayed by others) as a rural place. This has often been used to promote an image of tranquillity and quaintness, a sort of 'other-worldliness' in which Ireland is seen as something of a refuge from the modern world, a place in which traditional values of family, co-operation and friendliness abound. In the 1930s, the then Taoiseach (prime minister) Éamonn DeValera enunciated his vision of Ireland as a place of 'comely maidens dancing at the cross-roads'. This image of rusticity and youthful innocence (coupled with a devout Catholicism) was a metaphor for the nation. In part at least, this emphasis on rural imagery is associated with an opposition to an urban colonizer, England. To a considerable extent, urban areas in Ireland tended to be seen as those places most 'contaminated' by foreign influence. Following from this, rural Ireland has been presented as embodying the essence of the nation. The west of Ireland, with its rugged mountains, is seen as the 'real' Ireland, a place with a certain spiritual mystique which has been the inspiration for literature and other art forms (Figure 5.6). Artists such as Paul Henry (who was born and grew up in Belfast, the most industrialized city on the island) used the west coast and off-shore islands as the foci

Figure 5.6 County Galway, west of Ireland (Source: Author)

of their work. The west is seen to embody Irishness: mystical, romantic but also remote, desolate, de-populated (Nash 1993). All of these are images which themselves are seen to capture Ireland's existence at the edge of Europe and also its apparent social as well as geographic peripherality as a migrant society emptied of its people. These romanticized visions are those carried with them in their heads by those who migrate. While these images evoke a sense of nostalgia, they are also the predominant images of the island used in tourist brochures, on postcards, brochures and other material, designed to attract visitors to the unpopulated and remote landscapes. Such portrayals tend to play down the extent to which Ireland has become a much more urban society and one which, in an era of globalization, has ever more fully integrated into a wider world.

The north–south partition of the island in 1921 is frequently presented in terms of an 'unnatural' division of a small island. Northern Ireland comprises six of the nine counties of Ulster, the northernmost of Ireland's four historic provinces (Figure 5.7). In story and song, continued British occupation of the north is seen in terms of an unwarranted disruption to the island's territorial integrity. The four provinces are often referred to in song as the 'four green fields' with the ongoing territorial dispute over the north seen in terms of recapturing the fourth of these. A peace

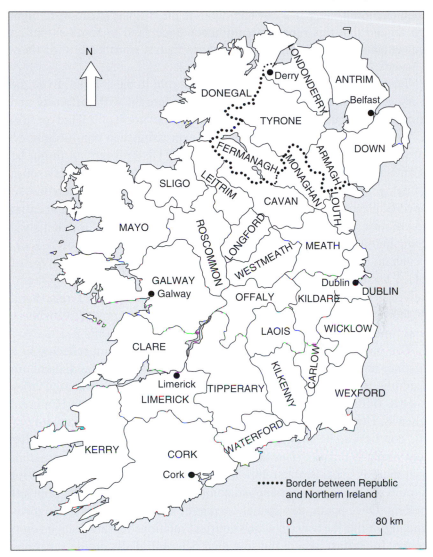

N

Derry

LONDONDERRY

ANTRIM

Belfast

DONEGAL

TYRONE

DOWN

FERMANAGH

ARMAGH

MONAGHAN

LOUTH

SLIGO

LEITRIM

CAVAN

MAYO

ROSCOMMON

LONGFORD

WESTMEATH

MEATH

GALWAY

Galway

OFFALY

Dublin

DUBLIN

KILDARE

LAOIS

WICKLOW

CLARE

KILKENNY

CARLOW

Limerick

TIPPERARY

WEXFORD

LIMERICK

KERRY

CORK

WATERFORD

Cork

•••••• Border between Republic
and Northern Ireland

0   80 km

Figure 5.7 Ireland

agreement in 1998 led to the creation of a power-sharing assembly
involving both Ulster unionists and Irish nationalists. While this endures
(despite some tensions), paramilitary activity by fringe groups periodically resurfaces, although the overall situation remains relatively stable
and peaceful. Nevertheless, the underlying issue of the future of the region
as Irish or British remains. Although the current resolution has been
accepted by the major factions on both sides, it is based on a recognition
of the ethno-national categories involved and, hence, an acceptance of

their territorial bases. In a way analogous to the Bosnian situation, it might be argued that this represents an acceptance (and to some extent an entrenchment) of divisions rather more than a transcending of them (Anderson 2008).

Territory continues to play a significant role in the conflict. Even the very act of naming the territorial entity is fraught with difficulty and ambiguity. Republicans tend to prefer the terms 'north of Ireland' or 'Six Counties' to 'Northern Ireland', thereby expressing their non-recognition of its legitimacy. Unionists prefer the term Ulster, partly, it might be surmised, because it avoids the word 'Ireland' and tends to confer a sense of being a place separate from the rest of the island. While references to specific places and the utilization of territorial strategies have been central to unionist tactics, it has been argued that a key problem for unionists is their inability to marshal an adequate set of place images of 'Ulster' (Graham 1994). Geographic imagery of Ireland as a particular type of place has been the preserve of republicans. Unionists, in seeing themselves as British, have dissociated themselves from images of Ireland and Irishness, but they equally do not fit neatly into notions of Britishness and cannot readily identify with place images of 'Britain'.

The political and sectarian fault lines in the north are reflected in a distinct territorialization. In urban areas in Northern Ireland, there are Catholic residential districts and Protestant districts. Essentially, cities such as Belfast have what are, effectively, Protestant ghettos and Catholic ghettos. However, it should be borne in mind that this ethnic reading of the conflict can be slightly misleading: not all Catholics are nationalist and similarly not all Protestants are unionist. Nevertheless, it is apparent that there is considerable religious segregation even if this cannot, and should not, be seen to correspond completely to people's wider political beliefs. Once again, this spatial segregation reflects power imbalances and highlights the ways in which political relationships are mapped onto space. In Belfast, where the so-called Shankill–Falls divide separates the two groups in the west of the city, a very high degree of segregation occurs, highlighted by the inappropriately named 'peace lines' – walls literally dividing roads in the area and designed to prevent confrontations between the two sides. Parts of the city are seen as out-of-bounds to one side because it is the other's territory and many lead lives which are spatially restricted to shops and leisure spaces within their own territory (Shirlow and Murtagh 2006). The territoriality of Belfast is such that a person's address is quite likely to reflect their place in a religious as well as a merely geographical sense. Similar forms of segregation pervade other urban areas (Figure 5.8).

Figure 5.8 Wall at boundary of Protestant Fountain district, Derry, Northern Ireland
(Source: Author)

Just as international borders have flags and other territorial markers to indicate their location, so also these segregated residential areas come complete with their own sets of boundary signifiers (see Box 5.1). Red, white and blue kerbstones indicate loyalist areas, while Irish flags and a variety of republican wall murals are highly visible symbols in nationalist areas. Murals may refer to specific events or places, some often include maps either of all of Ireland (republican) or of Ulster only (loyalist) while many commemorate events which occurred on the street or in near proximity to the actual site. The drawing of wall murals demonstrates the use of historical images in order to make contemporary political state- ments and it represents a micro-scale version of the larger conflict. While utilizing images of individuals or of historic events, the placing of these murals in public spaces, gable walls of houses or on other spaces which are easily seen serves a number of territorial functions. It uses a particular space to broadcast a political message, as with political graffiti (O'Reilly 1998). Some republican murals reflect wider international connections to places such as Cuba and South Africa (Figure 5.9). While a prime purpose of murals can be seen as sending a message to the muralists' own community and performing a politicizing function, they can also be seen as territorial markers: this space belongs to this tradition. Jarman

Figure 5.9 Irish republican wall mural, Derry, Northern Ireland (Source: Author)

(1998) suggests that initially murals were mainly for internal consumption but that in the 1990s some were clearly positioned to look outwards and be 'consumed' by others. They could thus be seen as signifying borders between different communities. The usage of murals, flags, kerb-painting and so on has the effect of making a 'Protestant' or 'Catholic area'. Once again territoriality allows a mobilizing process to take place.

Box 5.1 Marking territory

Within contested political spaces, the names given to areas or to streets reflects power relations or challenges to existing power (Berg and Vuolteenaho 2009). This can be seen in a variety of ways. At an official level, streets have often been named in honour of royalty or leading political or military figures or as a reminder of important battles. Significant political change is often accompanied by name changes. Thus, the collapse of communism and the dis-integration of the Soviet Union saw Leningrad revert to its earlier name St Petersburg, while in Montenegro, Titograd (named after the former Yugoslav communist leader) became Podgorica once more. Elsewhere in eastern Europe, street names were changed to reflect the changing political order. The creation of an independent Ireland in the 1920s saw Dublin's principal thoroughfare change from Sackville Street to O'Connell Street in honour of the nineteenth-century Irish nationalist Daniel O'Connell. This represented something of a reversal of the British colonial naming strategy which had imprinted colonial names or anglicized versions of Irish language names on the landscapes and streets of the island (Nash 2009). The defeat of the United States and the unification of Vietnam in the 1970s saw Saigon renamed Ho Chi Minh City after the former leader of Vietnamese nationalism.

While naming or re-naming places and streets is often an official recognition of a national claim, in conflict zones the unofficial changing of such things as street names are important symbolic ways in which territory is reclaimed, and the phenomenon was well known in black townships during the apartheid era in South Africa and also in Northern Ireland.

Parades have been a prominent feature of life in the north of Ireland for over 200 years. They are more strongly associated with the Protestant section of society where they are held to celebrate past events. The Orange Order is a Protestant organization which holds marches each summer in towns and villages throughout Northern Ireland (Bryan 2000). These marches have both religious and political meaning. They are seen as celebrations of a Protestant and British identity commemorating historical events, most notably the Battle of the Boyne in 1690 interpreted as a victory for the Protestant William of Orange over the Catholic King James. While many of these marches pass off peacefully, some have proved extremely controversial. A Parades Commission has been set up to adjudicate on whether some marches should take place and, if so, along what routes. Members and supporters of the Orange Order argue that marching is a basic right and a fundamental part of their Protestant heritage. From a republican perspective, these events are often seen as very visible expressions of loyalist triumphalism. The process of marching is not of course purely about history. As Bryan (2000) suggests, like many other displays of identity it serves a present-day purpose. It is seen to provide a link with the past, to assert a presence by that group and, hence, to legitimate their sense of identity and bolster contemporary political arguments.

An additional point here is that the use of the Irish language in public spaces has been significant. Language can be seen as an important signifier of identity quite often used by groups whose rights have been suppressed. While there is not a straightforward relationship between political inclinations and support for the Irish language, there tends to be a strong connection made between language usage and political allegiance (NicCraith 1995). The Irish language is not generally popular among Protestant communities, while in republican areas, some street names have been modified to incorporate the Irish version. Some shops have utilized bilingual signs and some republican murals incorporate Irish words. This has been made quite explicit by Gerry Adams, the leader of the major republican party Sinn Féin, who has argued that 'the restoration of our culture must be a crucial part of our political struggle and . . . the restoration of the Irish language must be a central part of the cultural struggle' (1995: 143).

The distinct spatiality of the conflict is highlighted in the experiences of people growing up in Belfast. In a memoir, Ciaran Carson, a poet from a Catholic background, quotes from the writer Robert Harbinson on his experience of this territoriality. From Harbinson's Protestant perspective, certain places were out of bounds:

> God ordained that even the Bog Meadows should end and had set a great hill at their limit, which we called the Mickeys' Mountain . . . In terms of miles the mountain was not far, and I always longed to explore it . . . But the mountain was inaccessible because to reach it we had to cross territory held by the Mickeys (note: a derogatory term for Catholics). Being children of the staunch Protestant quarter, to go near the Catholic idolators was more than we dared, for fear of having one of our members cut off.
>
> (Harbinson, cited in Carson 1998: 86)

Juxtaposing Harbinson's childhood memories with his own view from the other side, Carson reminisces about stepping over 'the slippery stepping-stones across the Blackstaff into the margins of enemy territory, which we approached with the same trepidation felt by Robert Harbinson, coming from the other side' (1998: 86). Both writers are expressing their sense of the city and the significance of the political and religious divide within it, which had a direct territorial impact on their childhoods (and those of many others). In various ways throughout parts of the city, the authority of the dominant political or ethno-nationalist group is either asserted or challenged.

Promised land

The Israeli state emerged in 1948 out of former Ottoman empire territory which had been under British mandate rule. The foundation of Israel has led to an as yet unresolved conflict over territorial control and national identity. Israel was created in order to provide a homeland for the Jewish diaspora scattered throughout the world. The creation of such a state was given added significance following the extermination of millions of Jews by the Nazis during World War II. Although there were other possible locations for a Jewish state, Israel was chosen, primarily because of its symbolic and historic connection with the Jewish people. The philosophy of Zionism portrayed it as essential that Jews acquired their historic territory, something which assumes huge importance for many Jews based on their religious history. In biblical terms, Israel is the 'promised land' and it contains within it the essential ingredients of the Jewish state. The problem that exists, however, centres on the reality that another recognizable national group, the Palestinians, also inhabit that space. Like all conflict zones, there are competing narratives with intense disagreement over both facts and the interpretation of them.

Originally, a United Nations agreement proposed the creation of twin states. Opposition to an Israeli state from neighbouring Arab countries led to conflict and, ultimately, to an Israeli state larger than that originally envisaged. The creation of Israel and the initial conflict led to the displacement of hundreds of thousands of Palestinians. The land vacated by those displaced Palestinians was given over to the settlers and new villages and towns were created. Many of the displaced Palestinians and their descendants have grown up in refugee camps in Lebanon and elsewhere. Not surprisingly, this led to the emergence of organizations such as the Palestinian Liberation Organization (PLO), and a range of others, dedicated to the achievement of a Palestinian state. The result has been a protracted and often violent conflict interspersed with numerous attempts at peace-making and the granting of limited Palestinian autonomy.

Following a 1967 war, Israel occupied the Palestinian West Bank (then part of Jordan) and Gaza Strip (until then, part of Egypt) and commenced a settlement construction policy aimed at making it their territory. Arab east Jerusalem experienced a similar process. Territorial appropriation and dispossession have been key elements in the conflict (Yiftachel and Ghanem 2005). Autonomous zones in the West Bank and Gaza Strip are now under Palestinian rule, though that is characterized by serious internal fractures (Figure 5.10). Israeli–Palestinian conflict continues in a region in which the idea of 'purified' space is made literal. Settlements are defined as 'Jewish' or 'Arab' or 'mixed'. The dispute has also had

Figure 5.10 Israel/Palestine

wider significance. For a time, the conflict became another element within the Cold War with US support for Israel mirrored by Soviet Union support for the Palestinians. In recent years, it fits, from an Israeli perspective, within the broader geopolitical narrative of struggles against terrorism. This intense and protracted dispute has also periodically led to serious tensions involving Arab states such as Syria, Jordan and Egypt and contributed to serious instability in Lebanon.

The actual utilization of territorial terminology has been part of the dispute. For a time it was illegal to utter the word 'Palestine' on the west bank or in Gaza (Said 1992). Indeed, as Said further points out, pro-Palestinian demonstrators on the streets of US cities during the 1960s and 1970s carried placards declaring 'Palestine' while their opponents' placards averred 'there is no Palestine'. Thus, the attachment of a name to the territory was itself an act of important political symbolism. Israel has pursued explicitly territorial policies, exemplified most obviously through the creation of Jewish settlements in the West Bank and the construction of a West Bank wall (ostensibly designed for security purposes) which cuts through Palestinian villages and farmland (Figures 5.11 and 5.12). Strict border controls are also imposed on the Palestinian territories. For some, a fundamental issue here is the 'structural elevation of Jewish over Arab citizens; the privileging of Jewish diaspora (and hence immigrants) over local Arab citizens; and the blurring of state borders, which allows West Bank Jewish settlements to continue to form a (de facto) part of Israel' (Yiftachel 2006: 5). While this conflict has proved intractable and is subject to diametrically opposed interpretations,

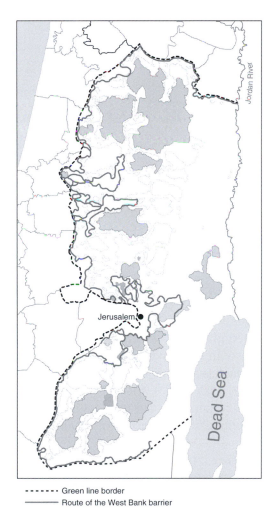

------- Green line border
———— Route of the West Bank barrier

Figure 5.11 Route of West Bank wall (Source: Ynhockey, Wikipedia)

Figure 5.12 West Bank barrier (Source: Etan J. Tal, Wikipedia)

it provides a classic example of a territorial dispute in which land is contested between two groups who both utilize historical and cultural arguments in support of their claims.

England's green and pleasant land

In a speech in 1993, the then British prime minister John Major spoke of those things which, in his view, exemplified Britain. It was a 'country of long shadows on county grounds, warm beer, invincible green suburbs, dog lovers and pools fillers' (cited in Billig 1995: 102). He spoke in a way which implied all Britons would agree with him. This evocation serves to illustrate the commonplace aspects of national identity, often associated with everyday spaces. Clearly the view is a very partial one, evoking rusticity and presenting a certain notion of timeless landscapes as symbols of the nation (to say nothing of the reflection of a male, rather than female, sense of affinity).

In any analysis of national imagery it is important to observe what is included (or excluded). Different strands can be identified in the territorial imagery normally associated with England. The first is a confusion between Englishness and Britishness. Major's vision of Britain highlights a central issue in that the exemplars used to illustrate Britishness might be seen as peculiarly English. Billig also cites Tony Blair, then in opposition but later to become prime minister of the UK, outlining his vision of Britain as a 'nation of tolerance, innovation and creativity . . . [with] a great history and culture' (1995: 105). This can be further seen as a subsuming of British identity into an English one. Considerable confusion exists in distinguishing English culture from British culture and in determining where English national identity ends and British national identity begins. For many people in England, there is a tendency to assume that English identity is synonymous with British identity. For people in Scotland and Wales, the distinction is clearer. Generally, they will see themselves as either Welsh/Scottish or British, or both. In the latter case, which of the identities, British or Welsh/Scottish, takes precedence will vary. However, for many English people, the distinction is unclear. Many of the symbols of Britain and Britishness are in fact emblematic of England – cricket, stiff upper lip, warm beer, village greens, rustic villages, gentle landscapes. The use of this imagery is designed to foster a sense of British national identity. But it is significant that, in so doing, there is the imposition of one of the subsumed identities, namely Englishness, as the signifier for the wider sense of Britishness. Englishness simply 'is'; it is the norm. To many this merely reinforces a sense of English cultural imperialism over Wales and Scotland. The territorial images of Britishness are those which might be seen as quintessentially English. Nevertheless, there is considerable difficulty in separating these. Furthermore, trying to establish when ideas of Englishness may have taken root is somewhat problematic and the subject of considerable debate (Kumar 2003).

The pre-occupation with such things as English sporting success (most notably in football) appears to predominate over a subordinate concern with the 'minor' nations within the kingdom. It might be argued that this reflects population levels and is compensated for by regional coverage on television and in other local media. However, this does not get beyond the fact that, at a 'national' level, England *is* the nation, in a way in which Scotland and Wales most definitely are *not*. This often gives rise to a certain bewilderment. English football fans cannot understand (or more likely understand but do not care) why their team is resented by many Welsh and Scottish people (and by many others as well). In part at least, these attitudes would appear to reflect a colonial past which continues to permeate the present.

Leaving aside the confusion between Englishness and Britishness, another contradiction emerges. English nation-building, like other national histories, is heavily dependent on antiquity. Somewhat perversely, many images of England, as with other countries, are rural and 'old' as distinct from urban and modern (Box 5.2). In part, this is bound up with a response to the processes of industrialization and urbanization. Constructions of the urban as 'evil' and contaminated have been mirrored by the emergence of the 'rural idyll', whereby all that is 'pure' and 'natural' is seen to be associated with the rural landscape and with rural life. These wholesome images are then taken as embodying the nation (Wiener 2004). In this way, the paintings of Constable are taken as representing Englishness and there is a long history of the use of idyllic landscape imagery to conjure up the nation so that, somewhat ironically, 'the pioneer of industrialization and the most urbanized country in the world is idealized in rural terms' (Taylor 1991: 151). While this linkage between idyllic rural images and national identity is by no means unique to England, Kumar (2003) argues that the elevation of rural landscapes as emblematic of England is characterized by the narrowness of the type of landscape – a southern English one of

---

Box 5.2 Arcadian images

In many countries, there is an image of rural areas as the 'real' country: the one which existed prior to the contaminating effects of urbanization. Linked to this is the elevation of agriculture as a wholesome and rewarding way of life. These images and beliefs have their roots in Arcadian notions of rural innocence, of people uncontaminated by external urban influences. In much of Europe, the latter part of the nineteenth century was characterized by rapid industrialization and urban growth. Cities were seen by many as smelly, dirty and 'unnatural' and this led to an anti-urban backlash. Many of those who migrated to the cities retained a nostalgic view of the countryside they had left. In Britain and elsewhere, versions of this rural idyll exist, although they are tempered in instances by the recognition of the existence of rural poverty and out-migration from rural areas. In Wales, Plaid Cymru originally encouraged moves 'back to the land' and gave agriculture an exalted status. The mountainous north and middle of the country were seen as the heartland of the Welsh nation: zones relatively uncontaminated by outside influences. In post-independence Ireland, there was an explicit invocation of the rural as the repository of something truly Irish. These rural images also serve to suggest a 'natural' state or a state of innocence. Within a nationalist discourse, these allegedly uncontaminated places symbolize the nation in its original and 'pure' form (Short 1991; Daniels 1993; Bunce 1994).

gently rolling hills. The construction of this rural idyll came at a time in the late nineteenth century when the majority of the population lived in urban areas. Williams (1973) argues there is an inverse relationship between the economic importance of the rural economy and the cultural value of rural ideas, suggesting that a nostalgia for a rapidly disappearing landscape and ways of life fed a particular socio-geographic representation of the nation (Figure 5.13).

Even when industrialization is being considered, particular places can be utilized in presenting a vision of the past. In this way, Ironbridge in Shropshire, has become a 'museum' of the industrial revolution with a number of different sites illustrating various aspects of England's status as the 'First Industrial Nation', a term which itself functions as a device to assert English national pre-eminence. In this celebration of industrialization, this one place has become a symbol for a wider vision of England's industrial heritage.

There are many sanitized versions of the nation's past which serve to generate feelings of warmth, nostalgia and goodness. A feeling is inculcated of how lucky people are to be English, to be proud members of a proud nation. Allied to symbols of monarchy, stately homes, alleged virtues of tolerance and 'fair play', territorial images and national characteristics create a view of England/Britain which is seen as comforting

Figure 5.13 'Idyllic' English village, Bibury, Cotswolds (Source: Author)

and re-assuring and, strangely, one which is seen as non-nationalistic. The dangers inherent in simplistic constructions of place may have negative impacts on those perceived to be different on the basis of ethnicity or other social criteria and who may find themselves subject to exclusionary practices and deemed to be 'out of place'. The construction of rural Britain as a relatively 'white space' reflects deeply embedded ideas associated with belonging, rurality and national identity (*Journal of Rural Studies* 2009). Similarly, the presence of gypsies and travellers in the British countryside has been a source of long-standing controversy centring on what are seen as the competing rights of settled inhabitants and nomadic populations (Holloway 2005). More broadly, ideas of British identity have led to various debates over citizenship and the ability, willingness or appropriateness of members of immigrant communities being regarded as British. The introduction of a citizenship test in 2005 for those wishing to acquire British nationality reflects a concern over the ability of newly arrived immigrants to integrate into British society, while simultaneously suggesting that some consensus can be achieved as to what constitutes British values, culture and identity.

Summary

This chapter has sought to uncover the geography implicit within nationalist narratives. It has demonstrated how a sense of history and geography combine to produce and re-produce a sense of national identity. Generalized territorial imagery and more place-specific images are utilized in order to create a sense of national well-being. The symbolic importance of place and the nature of territorial imagery have been demonstrated with particular reference to the former Yugoslavia, Ireland, Israel/Palestine and England/Britain. While such national geographies do not in themselves necessarily give rise to national chauvinism or overtly negative consequences, they can, in times of conflict, be utilized in support of territorial claims. The cases explored also indicate the use of territorial strategies, whether through the construction of settlements, the naming of streets, symbolic parades, the painting of murals, as a means of reinforcing a sense of the nation or as a means of resistance to a particular form of political control. In a variety of ways, territorial strategies are used not just as a means of exerting control over space, but to attain broader political goals. These contested spaces also reflect ideas of imagined communities and, of course, tell us something about imagined differences. While conflicts generally have a material base (connected with oppression or discrimination), ethnic differences tend to be emphasized and made 'real' so that each side creates and reproduces sets of assumptions about an essentialized 'other'.

Note

1   Lyrics available at www.lyricstime.com

Further reading

The works listed below deal with the importance of landscape imagery and its links to wider notions of nationhood, the links between place and national identity, and the emergence of national traditions in a number of different countries. Some material directly concerned with the examples used in the chapter is also included.

Boyce, D. G. (1982) *Nationalism in Ireland*, London: Croom Helm.

Bunce, M. (1994) *The Countryside Ideal. Anglo–American Images of Landscape*, London: Routledge.

Cosgrove, D. and Daniels, S. (eds) (1988) *The Iconography of Landscape*, Cambridge: Cambridge University Press.

Dahlman, C. (with T. Williams) (2010) 'Ethnic enclavization and state formation in Kosovo', *Geopolitics* 15(2): 406–430.

Daniels, S. (1993) *Fields of Vision. Landscape Imagery and National Identity in England and the United States*, Cambridge: Polity Press.

Gagnon, V. P. (2006) *The Myth of Ethnic War: Serbia and Croatia in the 1990s*, Ithaca: Cornell University Press.

Glenny, M. (1996) *The Fall of Yugoslavia. The Third Balkan War*, 3rd edn, Harmondsworth: Penguin.

Graham, B. (ed) (1997) *In Search of Ireland. A Cultural Geography*, London: Routledge.

Graham, B. and Howard, P. (eds) (2008) *The Ashgate Research Companion to Heritage and Identity*, Aldershot: Ashgate.

Graham, B., Ashworth, G. and Tunbridge, J. (2000) *A Geography of Heritage. Power, Culture and Economy*, London: Arnold.

Hobsbawm, E. (1990) *Nations and Nationalism since 1780. Programme, Myth, Reality*, Cambridge: Cambridge University Press, 1990.

Hobsbawm, E. (1998) *On History*, London: Abacus.

Hobsbawm, E.and Ranger, T. (eds) (1992) *The Invention of Tradition*, Cambridge: Cambridge University Press.

Johnson, N. (1995) 'Cast in stone: monuments, geography and nationalism', *Environment and Planning D: Society and Space* 13 (1): 51–65.

Kolstø, P. (ed.) (2005) *Myths and Boundaries in South-Eastern Europe*, London: Hurst & Company.

Kumar, K. *The Making of English National Identity,*, Cambridge: Cambridge University Press.

Lowenthal, D. (1998) *The Heritage Crusade and the Spoils of History*, Cambridge: Cambridge University Press.

Mazower, M. (2001) *The Balkans: From the End of Byzantium to the Present Day*, London: Phoenix.

Nairn, T. (1977) *The Break-up of Britain*, London: New Left Books.

Nairn, T. (1988) *The Enchanted Glass. Britain and its Monarchy*, London: Chandos.

Newman, D. (2002) 'The geopolitics of peacemaking in Israel-Palestine', *Political Geography*, 21 (5): 629–646.

NI Conflict archive. Available http://cain.ulst.ac.uk/.

Ruane, J. and Todd, J. (1996) *The Dynamics of Conflict in Northern Ireland. Power, Conflict and Emancipation*, Cambridge: Cambridge University Press.

Said, E. (1992) *The Question of Palestine*, (new edn), London: Vintage.

Shirlow, P. and Murtagh, B. (2006) *Belfast. Segregation, Violence and the City*, London: Pluto.

Short, J. R. (1991) *Imagined Country. Society, Culture and Environment*, London: Routledge.

Taylor, P. (1991) 'The English and their Englishness: "a curiously mysterious, elusive and little understood people"', *Scottish Geographical Magazine*, 107 (3): 146–161.

Timothy D. J. and Boyd, S. W. (2003) *Heritage Tourism*, London: Prentice Hall.

Toal. G. and Dahlman, C. (2011) *Bosnia Remade. Ethnic Cleansing and its Reversal*, Oxford: Oxford University Press.

White, G. W. (2000) *Nationalism and Territory. Constructing Group Identity in Southeastern Europe*, Lanham, MD: Rowman and Littlefield.

Yiftachel, O. (2006) *Ethnocracy. Land and Identity Politics in Israel/Palestine*, Philadelphia: University of Pennsylvania Press.

6 The future of the sovereign state

Due to various trends within the contemporary world the nation-state as we currently know it (and as it is misleadingly referred to) is seen to be under a range of pressures. The state system is portrayed in some discourses as increasingly unstable and potentially obsolete and the demise of the territorial state is seen as imminent by some observers. This is based on the view that it is becoming less and less relevant and is in the process of being superseded as a political and territorial formation. The state's predicted demise is attributable to a variety of factors which are seen to emanate from both above and below. Globalization (something of a catch-all term) is seen by many as heralding the end of the state as its borders are increasingly penetrated by a range of economic, social, cultural and political processes. The worldwide flows of goods, money, people and culture are seen to render national differences and state borders less and less relevant. As well as cross-border economic flows, it is argued that there is a range of other phenomena which are seen to transcend national boundaries. These include environmental concerns as well as the increasing importance of 'identity politics' and cultural diffusion, whereby there is assumed to be a decline in the significance attaching to national identity, states and borders. Such processes are said by some to lead to the annihilation of space and a consequent deterritorialization. Simultaneously, many states are faced with internal pressures (most obviously through secessionist movements) which threaten to fracture their stability. Issues of secession were touched on in the previous chapter but there are also states in which the lack of effective territorial control poses serious questions about their continued viability. This chapter investigates both these challenges to territorial sovereignty, issues of state failure and external global forces.

State failure

There is clear evidence of internal pressures on states, forcing them to break apart or to make territorial concessions to minority ethnic groups. The former Soviet Union and Yugoslavia provide graphic examples of this and have already been referred to in previous chapters. What was once the Soviet Union is now 15 independent republics. Some of these, most notably the Russian Federation, have themselves faced various secessionist pressures. In the 1990s, Chechen rebels endeavoured to create an independent state, a conflict that spread into neighbouring Dagestan and other parts of the Russian-ruled Caucasus. Elsewhere, other republics within the Russian Federation operate with varying degrees of autonomy. Russia's northern republics, historically incorporated into the Russian empire, subsequently the Soviet Union, and now the Russian Federation, exhibit elements of territorial sovereignty, coupled with keenly felt senses of ethno-national identity while continuing to (re)-construct economic, political and cultural relationships within the federation and beyond (Lynn and Fryer 1998). A number of the former elements of the Soviet Union have faced their own attempted secessions. As previously noted, Georgia, for example, has never fully controlled the breakaway regions of Abkhazia and South Ossetia. A war erupted over the latter in 2008 and the region remains beyond the control of the Tbilisi government.

Preston E. James (1969) talked about the effective state – that space under the direct control of the state. A breakdown of effective control leaves 'holes in the state' as with South Ossetia (see Muir 1997). Other examples of this include parts of Colombia deemed to be effectively controlled either by drug barons or by guerrilla armies, and parts of Somalia controlled by warlords. In a number of countries, particularly within the 'Third World', significant parts of the national territory may be beyond government control with militias under the direction of various groupings effectively policing these areas. Somalia, Sierra Leone and Liberia are good examples of where this has occurred. In extreme cases, it can be argued that these have become failed states where there is no effective governmental control coupled with a complete breakdown in state infrastructure.

These 'holes in the state' and more extensive state failure may result from a variety of factors. It follows from earlier discussions dealing with functionalist theories of the state that those states experiencing greater centrifugal forces will be more apt to be under considerable pressure in continuing to exist. Countries with significant ethnic or national divisions, linguistic differentiation or major regional inequalities might be seen as relatively unstable. In this way, some parts of the state may become virtually ungovernable by the central authority and may often come under the control of groups not aligned to the state as currently constituted. There is debate over levels of state weakness, failure and collapse. Activities such as maritime piracy associated with Somalia and elsewhere bring to the fore questions over the extent to which such activities thrive in the relative absence of a strong state (Hastings 2009). In places of intense instability, such as has happened in the neighbouring countries of Liberia, Sierra Leone and Guinea, suggestions of trans-state multiple-level governance rather than attempts to reconstitute unitary state power have been suggested (Sawyer 2004).

The view adopted by some observers is that some African states are virtually ungovernable with portions of territory controlled by various armed factions and state institutions effectively ceasing to function. Somalia is a classic example, where the lack of a proper functioning central state provides a rich breeding ground for warlords who are happy to use the territory as a 'gambling space' for their own desires, to borrow Wole Soyinka's (1996) phrase. Forrest (1988) distinguishes between 'hard' and 'soft' states. In part, the 'soft' states have never had stable mass parties; rather, they have had attempts to construct 'national' parties out of many varied factions and groupings. In some instances, these states may eventually 'collapse' (Muir 1997). While this idea of inherently unstable states is quite useful and appears to reflect political–territorial realities, there are two risks attaching to these views. The first is the danger of assuming that all African states, or indeed all 'Third World' states, are the same. This homogenizing view ignores the differences in history and geography of a complex part of the world. The second risk is the one of assuming the problem is intrinsically 'African': that it somehow reflects Africans' inability to govern or to look after themselves, that it represents the absence of a viable civic culture. Instead, the fault may well lie elsewhere.

> European avarice created a world of sovereign states, each one cast in the European legal mould. The people of the newly emancipated African colonies could either accept the world of the national sovereign actor or else be denied a place on the diplomatic stage. Not surprisingly, the political codes and institutions which had evolved slowly according to the shifting nuances of European life often failed to flourish when superimposed on African society. How unjust that Europe expects the world to believe that this is all somehow the Africans' fault.
>
> (Muir 1997: 212)

African states utilized European modular forms with, in many instances, quite disastrous consequences (Davidson 1992). In addition, it is clear from earlier discussions that unstable states are a political reality in Europe as well as in Africa.

In the face of secessionist or divisive pressures which threaten its stability and territorial integrity, there are a variety of responses which a state may employ. These responses may be either coercive or adaptive. In the former case, internal threats to stability may be dealt with harshly through the pursuit of a military strategy, as in the case of Serbia's (ultimately unsuccessful) handling of unrest in Kosovo. Overt military tactics are usually accompanied by harsh legal measures implemented through a punitive judicial system. The death penalty, lengthy prison sentences and institutionalized torture may all be elements within this oppressive response. The treatment of Kurdish separatist groups and the general suppression of Kurdish culture in Turkey is an example of a coercive reaction to secessionist tendencies. The obvious intention here is to quell resistance through overt suppression or jailing of dissidents and those seen to be fomenting strife, and to create an atmosphere whereby people are discouraged from taking action against the state.

However, there are more subtle, though perhaps equally effective, responses which fall short of outright coercion. The opinions, beliefs and aspirations of secessionist groups may be discredited or cast in a negative light through various forms of state propaganda or the dissemination of misinformation. In some cases, outright censorship of these views may be employed. For a time, legislation in both the UK and Ireland in the 1970s and 1980s made it illegal for television and radio to broadcast interviews with people who were members of Irish republican organizations. Whether overtly coercive or not, the intention underlying these methods is the same: to preserve the political and territorial integrity of the state. Through various mechanisms, an ideology of the state as the 'natural' unit is promoted and anything which destabilizes it is seen as 'unnatural'. This corresponds to Gramsci's ideas of the state retaining its hegemony through more subtle ideological rather than coercive means.

Alternatively, regimes may pursue a policy of accommodation or appeasement of secessionist claims as a mechanism for dissipating tensions. Forms of limited self-rule may be introduced in an attempt to head-off calls for complete secession. The decision by the British government to allow votes in Scotland and Wales on devolution in 1997 can be seen in the same way. While such decisions can be couched in terms of promoting democracy and encouraging a more people-centred approach, they can equally be interpreted as a means of countering the claims by Welsh and Scottish nationalists for total independence. In this way, the state is maintained through the somewhat paradoxical method of reducing some of its own powers (Bogdanor 1999).

The case of regional autonomy in Spain provides another example of this response. The Basque country and other regions were granted limited forms of self-rule in part at least to try to quell pressures for complete secession. Following the death of the dictator Franco in the 1970s, Spain's attempt to devise a constitution to reflect its internal divisions demonstrates the tension concerning limited autonomy within states. The 1978 Spanish constitution re-affirms the centrality of the Spanish state while simultaneously referring to both the Spanish nation and the nationalities and regions within it (Guibernau 1995). A similar tension occurs in Russia, whose 1993 constitution affirms the right of national self-determination but then expresses the inviolability of the Russian Federation, thereby preventing its component nations from seceding (G. Smith 1995).

While internal pressures weaken state stability and various states are threatened by forms of secessionist nationalism as identified in the previous chapter, these might be seen as threats to the existence and legitimacy of the *existing* state, or to the nature and extent of the state's territorial reach, but they are not necessarily suggestive of the demise of the state as a *concept*. Even where state failure is a very real threat or indeed a reality, it is often a case of factions competing for control of the state or parts of it and its resources. In other words, it is not the idea of the state that is in question, merely the right of the current one to govern particular territory. A more serious threat to the future of the state as a political–territorial formation is presented by external or trans-state pressures. It is to these that attention is now turned.

The end of the state?

In addition to internal pressures on states, there are also what can be seen as external threats, or pressures from above, to the viability of the state. These tend to result from the processes commonly placed under the umbrella of globalization. This refers to an increasingly interdependent and highly interconnected world where knowledge of other people and places is constantly available to us all and where there are regular interactions between people in one place and those in other places. Held *et al.* have referred to 'the widening, deepening and speeding up of worldwide interconnectedness in all aspects of contemporary social life, from the cultural to the criminal, the financial to the spiritual' (1999: 2). While debate revolves around the nature and extent of globalization, it is clear there is 'a process (or set of processes) which embodies a

transformation in the spatial organization of social relations and transactions . . . generating transcontinental or interregional flows and networks of activity, interaction, and the exercise of power' (Held *et al.* 1999:
16). The stretching, deepening, intensification and speeding up of social
relations across space and the development of a global communications
infrastructure have had profound impacts on states and their borders. In
effect, the world is being made a smaller place through global economic
transactions, cultural diffusion and social interaction facilitated through
telecommunications advances, information technology and electronic
media, combined with speedier, more frequent and cheaper air travel all
of which serve to bring previously 'remote' places closer to us. Some
geographers have likened these globalizing tendencies to time–space
compression; in an increasingly interdependent world, fashion and other
trends diffuse quickly around the globe and permeate political and
cultural boundaries quickly and easily (Harvey 1989; Massey 1994). It
is important to bear in mind that globalization is not a 'thing'; rather, it
refers to a set of processes which operate unevenly across both time and
space (Dicken 2011). In short, it is suggested that we live in a global
world in which there are expanding opportunities for interaction and in
which, as a consequence, territory and borders are of much diminished
significance. All of this suggests that the concepts of nation and state
are becoming increasingly irrelevant as we move towards the 'global
village' and an ever more cosmopolitan world in which questions of
nationality and place become less and less important and the bounded
spaces of the state are replaced by a system of global flows.

Various distinctive elements within the globalization process can be
identified. These can be subsumed under the broad headings of economic,
political, socio-cultural and environmental dimensions. However, it is
important to bear in mind that these are not discrete categories. Most
social scientists accept the interconnectedness of these elements: the
economy cannot be viewed in strict isolation from society, culture cannot
be detached from environment and so on. Indeed, one of the central
elements within debates surrounding globalization is the emphasis on
the inter-relationships of these elements. Nevertheless, for purposes of
explanation, these are briefly discussed below.

A global economy

It is within the economic arena that many of the most obvious signs of
globalization are manifested. There have been sizeable increases in the
extent of international trade and in the growth of foreign direct investment (Figure 6.1). Two key manifestations of globalization are increased

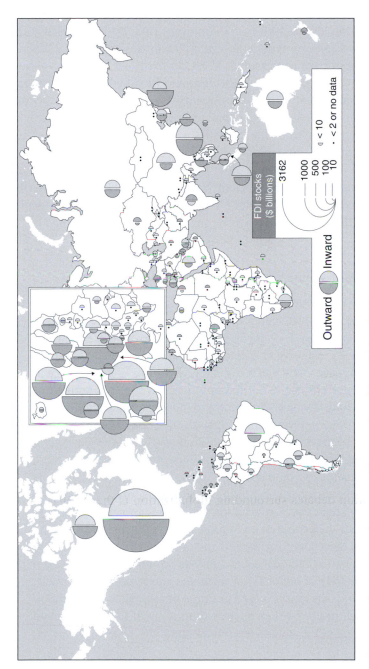

Figure 6.1 Foreign direct investment (Source: adapted from Dicken 2011)

inter-state economic co-operation and transnational corporations (TNCs), both of which call into question the sovereignty of the territorial state. Growing economic co-operation is reflected in the growth of supra-state institutions. These are regional amalgams of (usually) neighbouring countries engaged in formal economic or political alliances. The European Union represents the best-known example of this. Formed initially in the late 1950s as an economic union of six countries (Belgium, France, Italy, Luxembourg, the Netherlands, West Germany), it has now evolved into a more overtly political arrangement and membership has risen to 27 countries with a waiting list of applicants (Box 6.1). In some respects, the EU is beginning to take on the appearance of a 'super-state' in which power is taken away from the constituent members and is vested in Brussels. Similar trading blocs and economic agreements between neighbouring states exist within other parts of the world. These include the Association of South East Asian Nations (ASEAN) and the North American Free Trade Agreement (NAFTA); however, none is as well developed as the EU.

Box 6.1 European Union

The Treaty of Rome in 1957 established the European Economic Community (EEC) which has evolved into what is today known as the European Union (EU). Its original aims were broadly economic, centred on the creation of a common market between the six founding members: France, West Germany, Italy, Belgium, Luxembourg and the Netherlands. Since its formation, there have been a number of additions which have seen the union expand territorially to its present 27 member states (Figure 6.2). Just as there has been territorial expansion, so too there has been a political and economic deepening of the union. From its original focus on economic co-operation there is now an increasing emphasis on political co-operation. Attempts to move towards a harmonization of aspects of foreign policy and immigration policy are two examples of this. The spatial expansion and political deepening of the union have been resisted in some quarters. In most countries, there is an opposition to what some view as a diminution of national sovereignty as power is seen to be taken out of the hands of national governments and handed to Brussels. The EU is thus the most evolved example of inter-state co-operation – one which might be seen as providing a threat to the continued pre-eminence of the state as a political unit.

European integration proceeds both through supra-national institutions such as the European parliament and through inter-governmental co-operation over specific issues. Bodies such as the European Court of Justice, based in Luxembourg, now have a certain degree of power over individual states. In 1999, a single currency for the European Union (the Euro) was introduced. Despite this apparent usurping of state power by an external body, some such as Anthony D. Smith (1995) argue that national loyalties and national cultures will remain predominant, principally because there is no pan-European culture to replace them and, while the EU may take power from individual states, national loyalties seem unlikely to diminish significantly in the near future. While the EU rests on an idea of eliminating borders, it is busily engaged in hardening its external boundaries. The Spanish north African enclaves of Ceuta and Melilla epitomize this exclusionary tendency, ringed as they are by security fences designed to prevent African migrants accessing EU territory. Similarly, the EU eastward borders are increasingly portrayed as porous frontiers which need to be tightened, leaving the EU as a macro-level form of gated community (Walters 2004; van Houtum and Pijpers 2007).

A second economic threat to the sovereignty of the nation-state is the increasing economic power wielded by large businesses. Transnational corporations (TNCs) are large companies which have the power to operate in more than one country. By virtue of their size, many TNCs are in a position to exert considerable influence over government policy. They operate through a series of branch plants, the locations of which are determined by the relative differentials in wage rates, transportation costs, government grants, infrastructure, proximity to markets, etc. It is estimated that some 39,000 parent companies control 265,000 foreign affiliates (Dicken 2011). TNCs can take advantage of global sourcing opportunities, a highly geographical extensive set of markets, access to a wide variety of local labour markets and demonstrate a relatively footloose locational strategy. Around two-thirds of world exports of goods and services is accounted for by these TNCs and a considerable proportion of this is trade between branch plants of the same company. With many trillions of dollars of annual sales, these global corporations provide employment for many millions of people and their presence in particular countries is often seen as highly beneficial in terms of raising economic

Figure 6.2 European Union

output, employment levels and wage rates. In many countries, the amount of employment accounted for by branch plants of TNCs is extremely high. Many of the world's largest corporations have a considerably higher turnover than many 'Third World' countries. This, and other factors, has led to considerable criticisms of TNCs for operating effectively as neo-imperialist institutions (Klein 2000). Whatever the view taken on TNCs, they undoubtedly display something of a global reach which is relatively unhindered by territorial boundaries (Box 6.2).

Box 6.2 TNCs and the state: Shell in Ireland and Nigeria

The Shell oil company has operated in Nigeria since the 1930s and has extensive operations there, particularly in Ogoniland. Its activities here have been highly controversial and have engendered considerable opposition from local Ogoni people and from a range of activists concerned about Shell's environmental record. The relationship between state and company is clearly mutually beneficial. The picture for many of the state's citizens is, however, somewhat different. Criticisms of Shell came to a head with the judicial execution of the writer and activist Ken SaroWiwa – who had been very critical of Shell's activities – and eight others on charges of murder in 1995. It was felt by many that Shell could have prevented this and that the company was happy to profit from the oppressive policies of the state. In any event, Shell's power in Nigeria provides a useful illustration of the influence which large corporations are capable of wielding.

More recently, Shell has been involved as part of a consortium in a project to bring natural gas on-shore at Rossport in the west of Ireland, via a high-pressure pipeline, for processing. The plans, which enjoy the support of the Irish government, have met with popular local resistance with overlapping concerns related to land ownership (some local farmers have been faced with compulsory purchase orders by the state), health and control of resources. Local opposition is linked to broader national arguments within Ireland over who benefits from the exploitation of national resources and to environmental campaigns outside the country. In addition, links have been forged with Nigerian activists in Ogoniland and elsewhere in the Niger delta. In this example, issues of local and national politics intersect with questions of place, resources and environment (McCaughan 2008; Gilmartin 2009).

The growth of TNCs can be placed within the context of the rise of a neo-liberal economic orthodoxy which promotes the idea of limited state control of economic activities, encouraging cross-border flows of goods and allowing the penetration of domestic economies by foreign capital. Decades of light governmental regulation has led to the economic crisis unfolding since 2008 and afflicting many countries. Indeed, the global nature of this financial crisis is graphic evidence of the inter-connected nature of the global economy and the inability (or unwillingness) of individual states to control broader global economic forces. The collapse of the sub-prime mortgage market in the United States in 2008 has had a profound impact on economies across the globe. Greece, Portugal and Spain were particularly badly affected, while International Monetary

Fund intervention in Ireland has been portrayed by some as an effective loss or diminution of Irish sovereignty. It might be argued that a focus on borders and national economies is missing the point of a highly integrated global economic system. Nevertheless, it is important to bear in mind that a relatively small number of countries and economic regions retain a global dominance. Roughly one-third of TNCs have their headquarters in the USA, with a similar proportion in the EU and one-quarter in Japan.

Transnational politics

In the political sphere, the end of the Cold War was hailed in some quarters as a victory for social democracy ending, or at least diminishing, the ideological divisions and conflicts of the past. This argument was popularized through the publication of Francis Fukuyama's *The End of History* (1992). The collapse of communism and the subsequent advance of capitalism and democracy into eastern Europe were seen as homogenizing forces leading to the end of ideology and, hence, it was asserted, the end of major conflict. However, events since then, with the 9/11 attacks on the United States and the conflicts in Afghanistan, Iraq and elsewhere, have served to contradict Fukuyama's misleading and erroneous assumptions about the triumph of liberal democracy. Alongside all of this, the continued presence of nationalist conflicts (as discussed in previous chapters) serves to remind us that this is a far from settled world, politically or territorially.

At another level, the increasing evidence of political co-operation is taken as symptomatic of the decline of the importance of single-state action. In the moves towards political unity in Europe, or through the activities of the UN, there is seen to be a trajectory towards increasing political co-operation. Related to this, the evolution of systems aimed at harmonizing international law, the creation of various international legal protocols, the International Court of Justice in The Hague and a series of related developments can be seen as evidence of the creation of sets of supra-national instruments which operate across territorial boundaries and which internationalize politics. However, the power of these global structures is limited as evidenced, for example, through persistent ignoring of them by some countries, most notably the United States (Chomsky 2006). Similarly, the disregard in many countries for such matters as global standards of human rights suggest that state independence overrides attempts at establishing international benchmarks. Then US president George W. Bush's declaration of a 'war on terror' following the events of September 2001 and subsequent strategies amply illustrate the attempts by a small number of powers to shape politics at a more global level. The self-declared role of the USA as a sort of global policeman perfectly illustrates the uneven power relations within global politics. The hegemony of the United States and the 'west' sparks many debates over contemporary forms of imperialism but, however such hegemony might be regarded, it is certainly the case that other parts of the world are often treated as spaces to be controlled and monitored (Hardt and Negri 2000; Gregory 2004; Dalby 2007). Indeed, the USA has tended to view failed and collapsed states (referred to earlier) as places where dangers may lurk due to an apparent power vacuum and as potential centres for radical political activity. Failed states thus become threats in an era characterized by some as one of global terrorism (Elden 2007b). In addition, the neo-liberal agenda sees such places as Iraq increasingly secured through the use of private companies, rather than state security forces, thereby explic-itly linking economic imperatives with political strategies. In the rubble created by such interventions, opportunities emerge to re-shape the econ-omy and orient it to the advantage of Western business interests (Klein 2007). Harvey (2003) suggests that a territorial imperative is entwined with an economic one as spaces are controlled (or attempts are made to control them) in order to advance processes of capital accumulation.

In addition to these formalized versions of political integration and co-operation, there is a number of what might be described as alternative political visions which call into question the continuance of a world-state system. To some extent, these incorporate elements arising from below, within the borders of existing states, together with philosophies and visions which transcend state boundaries. Anarchist thought suggests a world free of formalized territorial divisions where borders do not exist. In a similar vein, variations of socialist thought point to alternative visions of the nature and functions of states. Perhaps more significant at the moment are social movements built around specific issues such as the environment or around various politics of identities, as discussed below. Such visions call into question the whole notion of borders and formalized territorial divisions. Alternative visions promote notions of co-operation rather than inter-state rivalry and they tend to envision a future whereby 'irrelevancies' of place of birth and national affiliation are seen as mean-ingless in a world dominated by a common humanity. The emergence of anti-capitalist protests in 2011 and entities such as the World Social Forum (and regional variants of this) can be seen as forms of protest at current trajectories but also as visions of pan-territorial approaches. They represent attempts to mobilize a global civil society built around issues of common concern and forging links between seemingly disparate groupings in various parts of the world.

Global culture

In the cultural sphere, people are wearing similar clothes, listening to the same types of music, reading the same literature (albeit, perhaps, in translation) and sharing similar values. The ubiquitous presence of CNN and other global media in hotel rooms around the world both reflects and sustains this process. It is further facilitated through Facebook, Twitter and related social media. Advances in mobile phone technology have allowed people from various social backgrounds and an array of cultural contexts to readily access information and to communicate more globally. While cultural differences persist, it is undoubtedly the case that a certain uniformity is becoming apparent. Step off a plane in Birmingham, Brussels or Bangkok and a McDonalds or BurgerKing is never too far away – the 'Golden Arches' of modern civilization beckon! People travelling with their laptops or computer notebooks are almost continually in touch with people elsewhere and the information super-highway brings the world to the palm of (almost) everyone's hand. The growing importance of a small number of languages, most notably English, as the international language of business and politics serves to further underline this process of cultural homogenization. Tourism, said to be the world's fastest growing industry, is another element which can be seen as a contributor to the flattening out of cultural differences (Urry 2002; Church and Coles 2007). Although there is an estimated 6,000 languages in the world, 60 per cent of them have fewer than 10,000 speakers and 25 per cent have fewer than 1,000 speakers. About two-thirds of the world's scientists write in English and 80 per cent of electronically stored information is in English (Murray 2006; Brower and Johnston 2007). In overall terms, the existence of highly networked transnational communities and ever more culturally diverse populations in some global cities might suggest that place-related national identities are of diminishing relevance (Vertovec 2009).

Linked to all of this is an increasing emphasis on other forms of identity. Groups mobilized around issues such as gay rights or women's rights elevate such questions of personal identity above concerns with national identity. In this way, cross-cultural solidarity and the growing significance of a politics of identity transcends political borders. This has resulted in considerably greater cultural mixing and the development of what might be seen as cross-national and cross-cultural solidarity in ways which undermine traditional borders.

Environmental concerns

There has been an upsurge of interest in the environment in recent decades characterized by a growing concern with the extent to which it is being damaged and abused. The environmental consequences of industrialization, increased car use, burning of fossil fuels and other activities seen as wasteful or harmful have contributed to the growth of an environmental movement and the development of green politics, with some electoral successes for green political parties. The growing problems of pollution, whether of water, air or land, are seen as ones which can only be resolved by international co-operation. Concerns over global warming and climate change have focused attention on the ways in which a world of sovereign states may act as an impediment to resolving environmental problems. Ideas of sustainable development and associated visions of a greener future emphasize the interconnectedness of the planet and stress the importance of a shared responsibility for its protection. The world does not belong to anyone but should be conserved and held in trust for future generations. Such a perspective sees state borders as irrelevant and points to the need to transcend them. Inter-governmental fora have been convened and agreements such as the 1997 Kyoto protocol on greenhouse gas emissions have been agreed. Many argue that governmental responses are much too limited in their scope and ineffective in their impact. A plethora of environmental campaign groups (some single-issue, others much more wide-ranging), reflecting a broad spectrum of environmental philosophies, lobby, campaign and protest in order to try to bring about more radical change. The global nature of many protest movements reflects not only the nature of the problems but also the internationalization of the responses. Various types of environmental legislation have resulted in different forms of territorializations such as the creation of national parks in many countries and a range of types of protected area where stricter planning controls and regulations exist (Figure 6.3).

Do states still matter?

Globalization has obvious implications for the world of nation-states. If globalization is occurring, then what is the future for independent political entities? In a nutshell, we live in an increasingly interdependent world where, it might be supposed, national boundaries are of ever-diminishing importance due to the processes outlined above. For some, such as Ohmae (1996), the state and national borders are both inefficient and increasingly irrelevant. Others, such as Friedman (2007), laud the flattening out of

Figure 6.3 Yorkshire Dales National Park, England (Source: Author)

global differences allegedly occurring and which diffuses useful technologies through the world. However, before we uncritically accept this
vision of a borderless world fuelled by increasing trans-border interactions, a number of qualifications need to be made here.

In essence, four key points can be made. The first is that globalization,
although a relatively new term, can be seen to incorporate processes
which have been occurring for a considerable period of time. Second,
globalization is not experienced by everyone, or at least not in the same
ways. Third, globalization is actively resisted in many diverse ways and,
finally, even within a globalizing world, it can be argued that states and
borders will continue to serve particular functions and will, as a
consequence, endure.

Nothing new

The phenomenon of an integrated world in which the fate of particular
places is linked to events in other, perhaps quite distant, places is not
simply a feature of the later twentieth and early twenty-first centuries.
States and regions have never been sealed containers. Trade and
commerce have always served to link people and places and the 'global
economy' is one which has evolved over time. Archaeological artefacts

point to extensive trading links extending well back in history and phenomena such as the central Asian silk route are evidence of established linkages between places well before the modern period. Colonialism was a project which linked places together – not necessarily to their mutual advantage. The 'triangular trade' involving European countries and their colonies, whereby raw materials and manufactured goods made their way to and from Africa and the Americas and people were shipped as slaves from Africa to North America and the Caribbean, represented a form of globalization whereby many diverse places were integrated into an increasingly global economy with a plethora of social and cultural implications (Figure 6.4). The British and Dutch East India Companies were the TNCs of their day. Within the broad realm of culture, there has been continual interchange of ideas, styles and modes of behaviour. Migration has been a long-standing social process and local cultures have continually evolved through processes of cultural interchange. In brief, it might be argued that globalization is occurring at a much faster rate than heretofore, but, although the term may be a relatively new one, the phenomena to which it refers are far from new.

In essence, the problem here is that the state has never been quite as sovereign as is often assumed. As Smith (2009) suggests, its monopoly

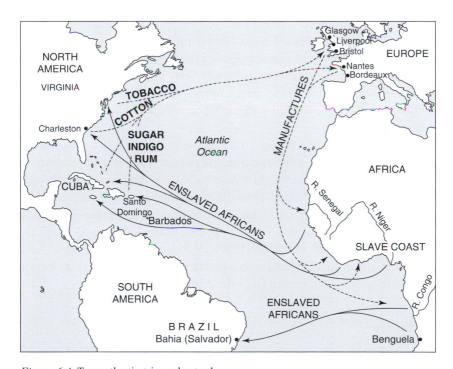

Figure 6.4 Transatlantic triangular trade

on violence, legitimacy and bureaucracy (Weber's formulation) was always partial and never complete. Strict sovereignty, as suggested earlier, is always a work in progress, never fully accomplished. Referring back to some sort of sovereign past when countries held complete control over all aspects of life within their borders is to idealize the history of the state as a political unit.

An unequal world

If globalization is not new, then neither is it universal; not everyone experiences it or at least not in the same ways. The world may be becoming a smaller place for academics and business people on transatlantic flights, as Massey (1994) argued some time ago, but it may be remaining much the same for those on the Atlantic or Pacific islands over which they fly, or for those sleeping under bridges in air-route hubs such as Bangkok. The stubborn persistence of glaring inequalities reflects the power of a global capitalist system to shut out the many while the few prosper. Similarly, not everyone lives in a world of CNN, Facebook and long-distance social interaction – only those with the time and money or social position which allows or encourages them to do so. Despite the undoubted rapid advances in the speed at which money, fashion, information and ideas travel around the world, and notwithstanding the evidence of vastly increased cross-border interaction, this picture of an ever more global world is not the complete story. While globalizing processes are occurring, there are many who remain locked into their own local worlds. Time and space are not shrinking for everyone. Viewed in this way, it is obvious that geography, territory and place continue to matter.

It should of course be pointed out that being locked into a local world does not mean that the outer world has no impact. First World companies can, through their decisions on where to locate, their wage levels and their employment conditions, determine the quality of life for people in places such as Vietnam and Thailand. Moreover, as Ferguson (2005) suggests, large corporations can hop from one location to another, bypassing chunks of 'unusable' territory to access those 'usable' bits which then become plugged into transnational networks while being simultaneously disconnected from their immediate hinterlands.

Resistance

Despite the obvious globalizing tendencies outlined above, and despite the view in some quarters that global processes are unstoppable forces, it is equally clear that these processes are constantly being resisted. Attempts at homogenization appear to always be met by efforts to resist and to assert difference. There are numerous examples of both overt and less obvious resistances to the flattening out of difference which is allegedly taking place. Within each of the arenas of globalization outlined above, we can see examples of resistance to globalizing processes. We might categorize some of these reactions as progressive and outward looking while others might be seen as more regressive and inward looking.

While TNCs undoubtedly wield enormous power, considerable opposition to their activities is evident particularly in many less developed economies where their power to operate without adequate environmental safeguards, to utilize non-union workforces and other practices are quite often resisted by local pressure groups. Their ability to dictate their own conditions and to open and close their operations in different countries is seen to lead to problems in terms of such matters as sustainable employment. The transitory nature of branch plants, derived from TNCs' ability to relocate relatively easily, is pointed to as evidence of the dangers attached to a dependency on such enterprises as a mechanism for economic growth or employment generation. In the view of many, the power wielded by TNCs is antithetical to the interests of many states and to sections of the population in particular countries. Disasters, such as that which occurred at the Union Carbide pesticide plant in Bhopal in India in 1984 when the release of a toxic gas led to thousands of deaths and many more maimed and permanently disabled, serve to highlight the dangers and provide a focus for opposition. A broad spectrum of protest groups have emerged, some issue based, some more broadly conceived, who in various ways endeavour to oppose the current trajectory of economic development.

Within the political sphere, the present ongoing process of European integration has taken on significant political dimensions and this has led to objections from certain quarters that national identity and sovereignty is being lost. In Britain, opposition to the continued economic and political integration of the EU coalesces around two broad perspectives. Those on the right of the political spectrum view current developments in terms of a loss of national sovereignty: a process which, it is suggested, is diminishing Britain as a nation and which is eroding its right to act in its own national interests. 'Outsiders' are increasingly seen to be interfering in matters deemed to be rightfully British. For many of those on the left of the political spectrum, closer integration carries the threat of a diminution of democracy. The EU is seen as a largely undemocratic body which is far too removed from the inhabitants of the member states and one driven by economic imperatives rather than democratic ones. In any event, power is seen to be taken away from individual member states and handed over to a central authority. Such moves have been, and continue to be, resisted in various quarters.

While many point to the end of the Cold War as ushering in a new era of global co-operation, it is readily apparent, as made clear in previous chapters, that national identity and fragmentation may well be the order of the day. The continuance of ethnic and nationalistic tensions in many parts of the world, most notably in eastern Europe, suggests that a world without borders is a long way off. The sub-state nationalism discussed in the previous chapter contradicts in many ways the idea of a global border-less world. Borders seem likely to continue to exist. Of course, this does not mean that the power of states remains the same. Diminutions in the extent of state sovereignty and effective alterations to the powers of states may well occur.

The growth of the heritage sector is a consequence of a range of factors, not least economic imperatives. But its success is certainly linked to the idea of preserving the local and the unique against modernizing and homogenizing forces. Re-assertions of interest in local languages, customs and cultural practices, although they may become closely bound up with tourism and the heritage industry (themselves symptomatic of a globalizing tendency), do nevertheless represent attempts at asserting or re-asserting a distinctive cultural identity and place-distinctiveness amidst the apparent homogenization of the twenty-first century (Storey 2010). It might be argued that this reflects a fear of placelessness but it has also fed into an increasing commodification of place (Figure 6.5).

Tightening borders

Another significant point must also be made. Although national borders have become increasingly more permeable, that does not mean that they have ceased to have relevance or that they are not needed. Even while recognizing the immense power wielded by some TNCs, it should be borne in mind that many of their activities are dependent on the existence of states. These include the propensity of TNCs to move from country to country, thus producing instability in employment levels and so on. Economies may become very heavily dependent on these corporations. In order to keep TNCs, countries may be prepared to alter or weaken employment or environmental legislation thus being effectively 'black-mailed' by these powerful non-governmental institutions. In this way, states may be said to be ceding control to supra-national private sector institutions. However, it is equally obvious that TNCs require the state

Figure 6.5 Heritage sign, Leominster, Herefordshire, England (Source: Author)

system. They could not engage in playing off one country against another if those countries did not exist in the first instance. The creation of export processing zones (EPZs) and the development of *maquiladora* along Mexico's border with the USA provide examples of capitalism's dependence on the state system. Even more fundamentally, production networks (while increasingly complex and transnational) are embedded in places within countries: goods are sourced, produced, assembled and sold to people in places (Dicken 2011) (Box 6.3).

Box 6.3 Export processing zones

The idea of spatial enclaves separate from the surrounding area is demonstrated in the existence of export processing zones (EPZs) (and related creations such as Special Economic Zones) in countries such as the Philippines and South Korea where 'normal' business regulations (such as environmental and labour laws and taxation regimes) are suspended or eased in order to encourage particular forms of production geared towards the export of goods. The number of EPZs has increased rapidly from 79 in 1975 to over

5,000, mainly in Asia, employing more than 40 million people. These have been subject to criticism for allowing large multinational companies to enjoy lax regulations in zones within developing economies (Klein 2000; Park 2005).

Similarly, the tactics employed by multinational companies engaged in extractive resource industries such as oil production often utilize forms of spatial enclaving. Angola is one of the world's leading oil exporters yet little of the benefits accruing from the extraction of this resource remain on Angolan territory, leaving it one of the world's poorest countries. Much of the extraction occurs off-shore, using imported labour and technology, and the profits, like the oil, flow out of the country (Ferguson 2005). More broadly, it is argued that non-contiguous privately secured economic enclaves effectively operate beyond the control of individual host states but instead are enmeshed in complex transnational networks.

A collapse in oil prices in the 1980s created a huge economic crisis in Mexico. This precipitated a shift in policy from domestic industrialization to the encouragement of foreign investment. Further economic integration resulted from the signing of NAFTA in 1994. One consequence of this policy orientation has been the creation of *maquiladora* or assembly plants of mainly US firms located in Mexico but just across the border from the USA in places such as Nogales, Ciudad Juarez, Nuevo Laredo and Matamoros. There are some 3,500 of these plants employing about one-quarter of Mexico's industrial workers. A very small proportion of the inputs to production are sourced in Mexico and the main activity is the assembling of products for export to the United States. While these contribute to the wider Mexican economy, there are relatively few backward or forward linkages to the local economy and they utilize relatively cheap labour while contributing to job losses in the United States where these activities previously took place.

Even within the EU, while borders are disappearing (although in 2011 Denmark moved to re-impose some border controls), an external perimeter fence is being erected making entry to the member states increasingly more difficult. There is an increasing emphasis on free movement of goods, capital and labour across the borders of the member states. However, this move towards equal internal access is accompanied by an increasing 'hardening' of the external borders of the EU. Entry of particular goods, such as certain foodstuffs, may be met with prohibitive common tariffs while moves towards harmonizing a common immigration policy have been gradually evolving. This has given rise to the notion of 'Fortress Europe' whereby internal freedom of movement is matched by an external perimeter fence which will become increasingly difficult to cross, particularly for those lacking skills and, by implication, those from poorer countries. As a consequence, one set of borders may be disappearing while another is being simultaneously reinforced. Ugarteche sums up this situation succinctly: 'globalization . . . simultaneously presents the younger generation with doors wide open to the world via cable TV, and with doors shut tight to impede illegal migration' (as we live) in 'a world which for some has no borders and for others has nothing but borders that cannot be crossed' (2000: 5). It might be argued that the evolution of regional super-states merely reflects a change in the present configuration of states, not a fundamental change in the idea of territorially based state control. A European border regime is developing in which a fear of the 'other' is emphasized while simultaneously cheap labour is sought and channelled in particular ways. Moreover, this border regime extends its tentacles beyond the boundaries of the EU itself so that the policing of Europe's borders is out-sourced to countries such as Morocco where walls separate Africa from the European enclaves of Ceuta and Melilla which are Spanish territory (Euskirchen *et al.* 2007). A capitalist system needs and sustains a border regime. Borders are a very necessary component within the system, and, though the level of differentiation and complexity may vary widely, borders, like states, may be reconfigured but are certainly not disappearing.

While the alternative political visions outlined earlier illustrate resistances to a state-centred world and suggest moves away from this framework, some cautionary words are necessary. The concerns enunciated above (environmental issues, gay rights, etc.) are more often than not articulated within a 'national' framework. In most countries, there are national groupings or national branches of international groups concerned with these issues – national branches of Amnesty International, Greenpeace and the like. They explicitly recognize the world of the nation-state and, while they do not necessarily conform to a state-centred way of thinking, they nevertheless see current territorial structures as a convenient organizational device. Even in attempts to proclaim the irrelevancy of territorial divisions, the methods used can be explicitly territorial.

Rather than superseding national identity, it may well be that questions of personal identity will continue to co-exist in a system of over-lapping identities rather than simply replacing national affiliations. Thus, one can think of oneself as female and/or homosexual and/or black but also Italian. Which identity takes priority at any one time will depend on the specific context in which the individual finds herself. In any event, whether or not national identities retain pre-eminence, it is abundantly obvious that there are strong resistances, many of them quite localized and territorially based, to processes of globalization and homogenization. Just as there are geographies of global integration so too there are geographies of resistance.

In conclusion, therefore, it can be argued that the world-state system is far from dead. Despite, perhaps because of, globalization, states and state boundaries seem likely to continue to function as significant elements in our everyday lives well into the future. In addition, there is no huge sign that national allegiances, as distinct from state power, are being significantly diminished.

Summary

This chapter has outlined two types of pressure said to be threatening the world of states as currently constituted. First, there are pressures from below seen to be associated with internal divisions within states and all or parts of the territory being subjected to contestation by various factions. Second, there are pressures from above largely associated with forces of globalization operating in the economic, political, socio-cultural and environmental spheres. While these various pressures undoubtedly exist, they will not necessarily lead to the demise of the state. Most internal pressures centre on the replacement of the current state with one or more new ones. Equally, many of the globalizing tendencies outlined above are actively resisted, suggesting that national perspectives and frameworks will continue to be of major importance. All of this suggests that rumours of the state's imminent demise have been greatly exaggerated and that place, and with it contestations over territorial control, will definitely continue to matter. Rather than a de-territorialization, it might be more meaningful to think in terms of territorializations at a range of different scales and which are constantly being transformed (Brighenti 2010a). Although there are clear internal tensions concerning the territorial extent of countries such as Iraq or Somalia, solutions are still sought within existing borders (Elden 2009). Resolutions to conflicts in various parts of the world continue to rest on a state-centric approach of either granting state-ness to groups or designing internal con-sociational agreements between competing groups (Anderson 2008). Sovereignty is highly contingent – more so in some places than others. Deals made between countries and corporate interests over African oil and other resources certainly reflect highly constrained ideas of sovereignty as conventionally understood (Carmody 2009). Ohmae's contention that the end of the state is nigh seems wide of the mark, and while we may exist in a world of flows, we continue to live our lives in territorially bound spaces.

Further reading

The material listed below provides insights into the various issues discussed in this chapter including internal pressures on states, the impacts of globalization, contemporary geopolitical debates and environmental concerns.

Ali, T. (2002) *The Clash of Fundamentalisms. Crusades, Jihads and Modernity*, London: Verso.

Carter, N. (2007) *The Politics of the Environment. Ideas, Activism, Policy* (2nd edn), Cambridge: Cambridge University Press.

Dicken, P. (2011) *Global Shift* (6th edn), London: Sage.

Dobson, A. (2006) *Green Political Thought* (4th edn), London: Routledge.

Gregory, D. and Pred, A. (eds) (2007) *Violent Geographies. Fear, Terror and Political Violence*, New York: Routledge.

Harvey, D. (2005) *The Enigma of Capital and the Crises of Capitalism*, London: Profile Books.

Harvey, D. (2005) *The New Imperialism*, Oxford: Oxford University Press.

Harvey, D. (2006) *Spaces of Global Capitalism. Towards a Theory of Uneven Geographical Development*, London: Verso.

Held, D. (1999) *Global Transformations. Politics, Economics and Culture*, Cambridge: Polity Press.

Held, D. (2004) *A Globalizing World? Culture, Economics, Politics* (2nd edn), London: Routledge,

Held, D. and McGrew, A. C. (2007) *Globalization Theory. Approaches and Controversies*, Cambridge: Polity Press.

Ingram, A. and Dodds, K. (eds) (2009) *Spaces of Security and Insecurity: Geographies of the War on Terror*, Aldershot: Ashgate.

Jennifer Milliken (ed.) (2003) *State Failure, Collapse and Reconstruction*, Malden: Blackwell.

Knox, P., Agnew, J. and McCarthy, L. (2008) *The Geography of the World Economy* (5th edn), London: Hodder.

Muir, R. (1997) *Political Geography. A New Introduction*, Basingstoke: Macmillan.

Murray, W. (2006) *Geographies of Globalization*, London: Routledge.

Pepper, D. (1996) *Modern Environmentalism: An Introduction*, London: Routledge.

Ritzer, G. (2007) *The McDonaldization of Society 5*, London: Pine Forge.

7   Territory, locality and community

So far in this book, the main concern has been with territories and territorial behaviour at a macro-scale centred on discussions of nationalism, the state and inter-state territorial relations. However, as indicated earlier, territorial behaviour and territorial strategies can be seen to operate at much smaller spatial scales. Some aspects of this, those surrounding sub-state nationalisms, have been addressed in previous chapters. This chapter, and the one which follows, focus on the creation of territorial entities and the utilization of territorial strategies at a sub-state level. In this chapter, the focus is on what are largely formalized territorial divisions beneath the level of the state. The following chapter deals with what might be seen as informal territories of a somewhat more ephemeral nature, focusing on issues such as racialized space and gendered space.

States also utilize internal territorializations. These may be internal federal units or, even in more centralized states, the use of local or regional forms of government (counties in the UK). Other territorializations may reflect specific forms of government policy as in the creation of designated spaces for purposes of landscape conservation (such as national parks) or forms of regional planning (such as designated areas or enterprise zones). In addition, the electoral systems of most countries are based on spatial units (constituencies) such that elected representatives have a territorial base representing the residents of bounded areas (Box 7.1).

Box 7.1 Constituency boundaries and gerrymandering

Within the formal political system, another version of territorialization is utilized. This is the construction of geographically based political units for electoral purposes. Generally, these are known as constituencies: areas in which prospective members of parliament stand. While constituencies may not have the same resonance as local authority boundaries and the like, they do provide an example of a territorial strategy in the realm of formal politics. Politicians may behave in a very clientelist way whereby they perform 'favours' for constituents in order to ensure re-election. They may also be keen to lobby on behalf of developments viewed favourably by the majority of their constituents. This might be reflected in support of particular proposals aimed at job creation in certain areas, or it might be reflected in opposition to schemes seen as detrimental to the local environment. This clientelist approach to politics can apply at all levels within the electoral hierarchy from MPs down to parish councillors and it highlights again the connections between place and politics and how territorial constructions can be utilized in order to bring about a desired outcome (whether from the point of view of constituents or the politician concerned).

Ostensibly, constituency boundaries are meant to reflect objective criteria related to population size and demographic change. However, the construction of constituency boundaries is not necessarily a neutral process. It can be manipulated in ways that work to the advantage of particular parties or to the disadvantage of others. Gerrymandering refers to the deliberate manipulation of electoral boundaries so as to gain an advantage through a malapportionment of voters between constituencies. It usually involves the construction of electoral boundaries such that the votes of the opposition are effectively wasted or the advantage accruing from them is minimized while the manipulators' votes are maximized. The term was coined following boundary alterations by Elbridge Gerry, the Republican governor of Massachusetts, in 1810 which worked to the benefit of his party. The resultant electoral district was shaped like a salamander, giving rise to the term 'gerrymander'.

The initial concern in this chapter is with internal territorial divisions which governments or their agencies use in order to 'manage' the state or to govern more easily. This is generally done through the creation of sub-state geographic units within which local or regional bodies have some form of control. These internal territorial structures may range from those which are relatively powerless and weakly developed, as in the case of highly centralized states, to those which have a considerable degree of autonomy, as in the case of federated or highly devolved states. This chapter initially considers these formal state territorial divisions. In addition to these divisions, there is a range of quasi-state or semi-state bodies with particular responsibilities in relation to their designated areas. These include regionally based health authorities, police forces and the like. Together with local and regional governments, these various agencies can be said to constitute the local state: a set of social and institutional relationships existing at a local level (Box 7.2).

Box 7.2 Policing space

The geography of policing has begun to attract some research interest since Fyfe drew attention to it in 1991. One elementary way of seeing this is the creation of policing divisions with responsibility for particular regions, such as West Mercia police in the UK and so on. Linked to this are ideas of specialized spaces in which special units operate such as harbours and airports, where particular concerns linked to immigration, smuggling and security are seen as priorities. The manner in which policing is carried out also reflects ideas of territory and territorial control ranging from police having particular 'beats' or zones of responsibility to the spatial tactics employed by police; for example, funnelling street marches and protests along agreed routes. Recent protests in London (in particular, student demonstrations in 2010) saw the police detain protestors in confined spaces – a controversial policing strategy known colloquially as 'kettling'. The increasing array of non-state actors engaged in forms of policing also brings with it a series of additional territorializations with private security firms patrolling office blocks and shopping malls. These create various tensions, including attempts to prevent people taking photographs in certain spaces. The advent of citizen-based groups such as Neighbourhood Watch also rest on notions of local territories with people encouraged to report suspicious behaviour. Current UK government strategies appear designed to deepen trends towards greater community control and the broadening of policing partnership arrangements.

Policing is about more than enforcement, it is also about the regulation and endorsement of specific values and moral codes (Mawby and Yarwood 2011); so in addition to territorializations associated with the policing of space, there are those which endeavour to exclude individuals or particular forms of behaviour from certain spaces. In the United Kingdom, anti-social behaviour orders (ASBOs) and related instruments provide a controversial example of bringing a territorial approach to dealing with particular behaviours that are seen to constitute a nuisance. They were introduced in 1998 and they can be used to prevent specified individuals from being in particular places, streets, etc. or to prevent individuals from repeatedly engaging in specified behaviours such as aggressive begging, disturbing neighbours, or drunkenness in public places. They have proved controversial, seen by some as the criminalization of certain social behaviours (Squires 2008). What they clearly do is reinforce the idea that certain forms of behavior are not allowed in public space and they can be punished with territorial exclusion. What is also clear is that they tend to be imposed on people in certain types of place, mainly areas of public housing in UK cities (Painter 2006). Similar so-called 'bubble' laws in the USA also function to exclude people behaving in certain ways from specific places (Mitchell 2005). Regardless of our view of the rights and wrongs of such moves, and whether they might be deemed successful even on their own terms, they raise important geographical questions related to what is meant by 'public', who is allowed to be where and how people are expected to behave (Cameron 2007).

Within countries such as the United Kingdom, the programme of 'rolling back the state', driven by a neo-liberal economic agenda, has led to an increasing reliance on the private sector to provide services previously seen as the preserve of the state or its regional organs. The increasing numbers of local service providers means that the local state is becoming ever more fragmented. Associated with these recent changes in how services are provided is an increasing emphasis on the role of local territorially based communities. Community-based organizations deemed to represent the interests of residents within defined localities are increasingly being drawn into the formal political arena where they may act as important agents in creating and sustaining a territorial identity. All of this means that these various agents (commercial and voluntary) are becoming part of new territorial arrangements in which the division between the public and private sectors is becoming increasingly blurred, leading to a consideration of modes of governance rather than government. Central to this chapter is a concern, not so much with documenting the specifics of territorial arrangements, but rather emphasizing the manner in which space is used in order to administer and control. As with the examples in previous chapters, these internal territorial arrangements represent a spatial expression of power: a means through which the state or its agencies manage their affairs.

Sub-state territorializations

Within most countries, there are internal sub-state political–territorial divisions. These are geographical sub-units which enjoy some limited powers over their own affairs. The extent to which states devolve power to local territorial formations is quite spatially variable. As already suggested, some states have highly devolved political systems while others are highly centralized. Moreover, within individual countries, the extent of autonomy accorded to sub-state territorial units alters over time.

At a basic level, a differentiation can be made between centralized states and federal states. In the former, power resides almost exclusively with central government with only a minor role, if any, for regional or local bodies; in the latter, various powers may be devolved to regional or provincial governments. In a federal state, such as the United States or Germany, regional governments have a considerable degree of autonomy and can pass local laws. In contrast, highly centralized states have a system in which the degree of local devolution may be quite minimal. Obviously, many countries adopt political systems which exist on a spectrum between highly centralized and highly devolved. In the following sections, a brief discussion of federal states and local government is provided. While these spatial arrangements exist primarily for administrative purposes, it needs to be remembered that territorial divisions at this scale, just as with those referred to in previous chapters, may also inculcate a particular sense of place identity. The recent trend of rolling back the state in some Western democracies is briefly discussed, followed by a theoretical appraisal of decentralization.

Federated states

The most devolved political–territorial systems are in federal states in which a significant amount of decision-making is carried out at a local level. Federalism is a political arrangement often favoured in multi-ethnic states. The ostensible aim is to ensure a reasonable balance between national and regional interests and (usually) between the various ethnic groups living within the confines of the state. Federal states can be seen, in many respects, as offering a practical response to tensions which might otherwise undermine the state's continued existence. Ethno-regional groupings are accorded power over their region within an overall federal system. In some instances, complex systems have evolved to try to deal with a multiplicity of ethnic, religious and/or regional tensions, as in India and Nigeria. In the latter case, more and more federal units have been created since independence and, as noted in Chapter 4, a sense of a unified Nigerian identity has been undermined by deep regional tensions. Some countries favour relatively large federal units, as in India, others favour very small units as in the Swiss *cantons*.

An obvious example of a federated structure is the Russian Federation, which consists of a range of what can be seen as virtually separate countries bound together under the overarching authority of the Moscow government. Within the Russian Federation, there is a complex arrangement of over 80 different types of federal unit, some of which enjoy greater autonomy than others (Figure 7.1). Ultimately, what makes these territorial units something less than fully independent states, in the sense indicated in Chapter 3, is their lack of power in the arena of foreign policy and their consequent lack of complete sovereignty over their territory: they remain subject to Moscow and have no international recognition as independent sovereign entities, though some may aspire to independence and/or behave in a relatively autonomous manner.

The United States provides a further example of a federal structure. Here, each state has its own governor and legislature and can implement its own laws on issues such as the death penalty and homosexuality. While each state has a range of powers, they all remain subject to the overarching political institutions of the federal government based in Washington and to which each state sends political representatives. Individual states do not have sovereignty – that resides with Washington. Spain is another example of a country employing a federal or quasi-federal mode of government; its 17 provinces enjoy limited self-rule within the confines of the Spanish state in a somewhat asymmetrical system where the Basque country and Catalonia enjoy considerably more autonomy than the other provinces. As indicated in the previous chapter, this has not

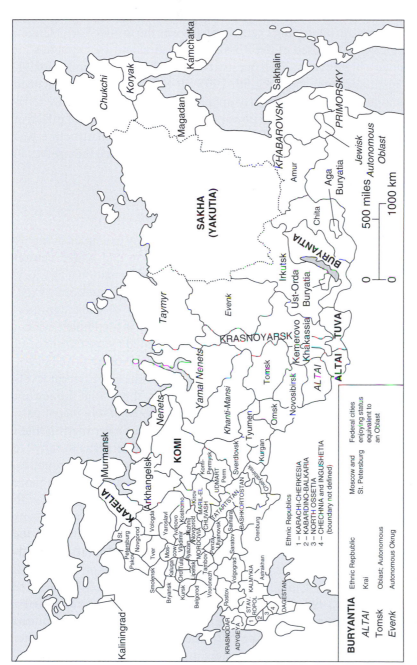

Figure 7.1 Russian Federation

prevented calls for secession on the part of certain regions, most notably the Basque country.

The name of a country may suggest a particular form of territorial administration. However, there are states whose names suggest a highly devolved structure but where, in practice, power is tightly concentrated in the centre. An interesting example of this was the former Soviet Union. While it was composed of the Russian Federation and 14 'republics', it remained a highly centralized one-party state: a type of 'federal colonialism' instituted from above (Smith 1995). It follows that there is a need to distinguish between what may appear to be a highly internally territorialized system of devolved power and the reality in terms of effective devolution of power. Yugoslavia's underpinning federal structure was insufficient ultimately to bind together many who defined themselves in different ethno-national terms.

Local government

In contrast to truly federated systems, there are many examples of highly centralized states where there is little devolution and where central government exercises tight control over all parts of the national territory. However, even in these cases, internal territorial subdivisions often exist. States still need mechanisms through which they can manage their sovereign territory. In the United Kingdom, there is a complex system of local government, a situation further complicated by the existence of different systems in England, Wales, Scotland and Northern Ireland.

Local government in the United Kingdom operates at different levels within a spatial hierarchy. Two major changes in the spatial framework of local government have taken place in the comparatively recent past. In 1973, many 'old' counties were replaced with newer metropolitan counties while others, such as Herefordshire and Worcestershire, were amalgamated. In 1998, there was a partial reversion to the older counties. Beneath the county councils are sets of district councils and throughout much of Britain a two-tier system of local government has operated with responsibilities split between county and district councils. Basically, items such as education, transport and social services tend to fall within the ambit of county councils, while concerns such as housing, refuse collection, elements of local planning and recreational provision tend to fall within the remit of district councils. This situation has been made more complex by the 1998 changes which have resulted in the creation of a number of unitary authorities in some counties (such as the re-constituted Herefordshire) which combine the functions of county and district councils. In others, such as Worcestershire, the district tier remains.

Beneath the county and/or district councils lies another tier. Parish councils (of which there are over 8,000 in England) have a right to be consulted by district or county councils over issues affecting their area but they have very few powers of provision, although they can provide or improve things such as village halls, car parks, camping sites and footpaths at their discretion (Bennett 2006). If the situation with regard to local government is complex with some areas of apparently over-lapping responsibility between the tiers, it is further complicated by the fact that numerous quasi-governmental non-elected bodies have responsibility for certain functions. These include local transportation authorities, health authorities and urban development corporations. In addition, in the late 1990s and early 2000s the UK also experimented with regional development agencies tasked with encouraging economic regeneration at a regional level. These are now in the process of being abolished. The trend towards the privatization of public utilities in many countries (referred to earlier) means that many previously publicly provided services are now provided by private companies, rather than state-run utilities. These companies tend to operate on a territorial basis. Within the UK, examples include Severn Trent Water in the English midlands and the electricity company Swalec in south Wales.

While the system of local government in the UK is largely based on historic divisions (partly ecclesiastical) of long-standing, the French system of *départements* was constructed for administrative reasons after the revolution in the eighteenth century. At the time, this was an historic attempt to create a 'rational' system of local government (Anderson 1996). The *départements* were instituted in order to form the territorial basis from which to administer the new republic. Although there have been changes in the degrees of power wielded at this level, the divisions themselves have endured to the present (Figure 7.2).

Theories of decentralization

Local government can be placed within the same theoretical structure as our examination of the state in Chapter 3. Viewed from a liberal–pluralist perspective, forms of local administration can be seen as a set of mechanisms through which local needs can be ascertained and then met. In this way, it might be argued that local government serves to bring about an acceptable solution to local problems. It might also be seen as facilitating the resolution of conflicting views at a local scale. The agencies of local government might be seen as providing a mediating filter between national interests and local interests and also between competing interest groups at a local level. In addition, local government can be viewed as

Figure 7.2 French *départements*

a means of evening out the spatial blindness of overarching national policies through making them more sensitive to particular geographical areas and to local circumstance. Clearly, such internal territorial arrangements do render it easier to deliver services at a local level, while on the political front, local or regional government (at whatever scale of devolution) can be seen to represent an attempt to deepen democracy by bringing the political process down to a local scale: it might be seen as furthering local control over particular issues. From this liberal–pluralist standpoint, the local state is seen as a means through which the central state can be more responsive to local needs while providing a set of mechanisms through which greater public participation can be achieved. This also implies that such arrangements allow the state to operate in a more efficient manner.

To subscribe to such a view is to see local government and regional autonomy in purely functionalist terms: as merely being a mechanism for territorial management. However, it can be argued that such systems have a deeper significance than this. It is easy to regard devolved systems of power as a genuine attempt to democratize decision-making and to make regions more autonomous, but there may be a number of other factors leading to the creation of sub-state territorial units with some degree of autonomy. Following from theories of the state discussed earlier, it could be argued, from a structuralist perspective, that the territorial sub-divisions are a mechanism through which the state can better organize itself. In order to retain its territorial integrity, the state needs to maintain its hegemony over all its geographical sub-areas. Often this is best done through sub-state structures. Through internal territorial arrangements the state can ensure its legitimacy over all its geographic space. Even in highly devolved states, it can be argued that the ultimate aim is not greater democracy or sensitivity to local needs, but ensuring internal state hegemony. For example, devolution of power may serve to bring about the elimination or reduction of separatist pressures. Viewed in this way, the local state serves merely to bolster the central state and, in so doing, it enhances its ability to reproduce the existing political system and to further the inter-linked processes of capital accumulation, labour reproduction and the maintenance of social order (through local policing, etc.). By being seen to involve local decision-making, the existence of local government also serves to further legitimate the state by promoting the veneer of greater democracy. This is not to argue that democratization is not occurring, rather it is to suggest that this is an 'accidental' outcome as opposed to the underlying purpose.

In Spain, as indicated earlier, there are 17 autonomous regions dating from the 1970s, but based to a large extent on historical territorial formations. The regions were granted autonomy in the aftermath of the rule of the dictator General Franco, who ran a highly centralized corporatist state in which attempts at local autonomy and expressions of cultural difference within Spain were suppressed. As previously suggested, autonomy for the Basque country has been a response to demands for full political independence by militant Basque separatists. Spanish political regionalism might be interpreted as an example of devolution practised in order to try to maintain and strengthen the state, not to weaken it. In this way, regional autonomy for the Basque country, Catalonia and elsewhere might be seen as attempts to stave off the disintegration of the Spanish state through the promotion of limited autonomy and the acknowledgement of cultural differences. Federalism or other highly devolved structures can be seen as political–territorial strategies implemented in an attempt to reconcile the competing centrifugal and centripetal forces to which states are subject, as suggested in Chapter 3. Franco's highly centralized state was envisaged as a means of suppressing separatist tendencies while the granting of significant autonomy can be seen as a different (and considerably less oppressive) way of dealing with these forces, while at the same time retaining the regions as integral parts of Spain. Continued pressure for independence for the Basque country and Catalonia and less overt separatist tendencies in regions such as Galicia tend to suggest that such federalizing policies are not entirely successful (Figure 7.3).

The United Kingdom has implemented a degree of regional autonomy with Scottish and Welsh assemblies following referenda held in 1997. In a slightly different context, Northern Ireland has its own elected assembly, part of an overall 'peace strategy' designed to end years of militant political conflict (discussed in Chapter 5). While it can be argued that these represent genuine attempts at enhancing regional democracy within the UK, critics might argue that they can be interpreted as efforts to maintain the union through offering limited devolved powers in order to stave off complete separation and, consequently, the dissolution of the existing state (Bogdanor 1999). Viewed in this way, such devolutionary strategies might be seen as an attempt at maintaining state hegemony rather than a weakening of it. The electoral success of the Scottish National Party in 2011 suggests that this policy is of limited success.

In short, while devolution, federalism and related strategies may well generate more efficient service delivery systems, and while they may enhance local democracy (although neither of these are necessary

Figure 7.3 Political graffiti, Galicia, Spain (Source: Author)

outcomes), they are primarily forms of state territorial management. In order to maintain its territorial hegemony, the state has two main methods. The first is a policy of assimilation, the other is one of devolution. The assimilation argument is that all regions should be treated more or less equally with no effective concessions to local sensitivities (as in Franco's Spain). The devolutionist pathway is one where power is devolved to constituent territories. Ultimately, both approaches are designed to prevent territorial secession and to maintain state control over all regions.

Rolling back the state: from government to governance

The rise of 'new right' politics in the UK and many other Western democracies since the late 1970s has resulted in a change in the way in which the state is viewed, with consequent implications for the role of local government. The key shift has been from provision to enablement, whereby local authorities are now operating more as facilitators of service provision rather than delivering the services themselves. This shift reflects a free market ideology aimed at the minimization of state intervention. This has involved the de-nationalization of state companies and services and their subsequent privatization. Associated with this has been an increased tendency towards state agencies contracting-out services. The policy of 'rolling back the state' might be said to be, in part at least, driven by the idea of saving government money through placing responsibility for service delivery in the hands of the private sector. However, while economic arguments were often used, there was undoubtedly a strong ideological component to this strategy. State intervention in the area of service provision (such as water, electricity and gas) was viewed in negative terms and there was a strong emphasis on individuality in preference to collectivism. The state was viewed as an entity whose authority needed to be reduced and whose influence on everyday life should be minimized, precipitating a shift in emphasis from the state to civil society. By and large, this ideology, based on the promotion of entrepreneurialism rather than the straightforward management of services, has remained in place.

It followed from the arguments outlined above that local government should be considerably reduced. Within the United Kingdom, the Conservative programme instituted by Margaret Thatcher in the 1980s, and continued by subsequent governments, saw an erosion of local government power combined with the promotion of a different ethos. The roles of both central and local government as service providers were considerably reduced with a range of quasi-state or non-state institutions filling the resultant vacuum. Thus, housing associations, health trusts and the like have taken over powers of allocation and provision of specified services. The market place rather than the ballot box has become the medium of popular expression as the private sector takes over the role traditionally reserved for arms of the national or local state.

These moves have been heavily criticized by many. They are seen as representing a diminution of democracy by transferring powers from elected local authorities to unelected quasi-autonomous non-governmental organizations, or quangos as they are commonly known. Places are being governed via a range of extra-state institutions. While the democratic nature of such moves may be questioned, from a theoretical level they have sparked a consideration of how the state operates. This is further compounded by an increased emphasis on voluntary and community-based activity, to which we return later in this chapter.

Administrative divisions and a sense of place

The above discussion has presented territorial strategies in broadly functional terms. The basic view propounded is that internal territorialization is a mechanism for the retention of political control. Despite this, however, the significance attached to these territorial formations should be borne in mind. As was observed in relation to national identity and state formation, the erection of boundaries leads to the creation and sustainment of territorial identities. If particular territorial formations have a long history, or are assumed to have, then people may have a strong identity with their region or at least with some notion of their own locality. Territorial sub-divisions within states may have long historical origins. Such is the case with counties in England. The re-organization of local government in 1973 led to the creation of many new counties which had no historic basis. In part, this re-organization was designed to create a number of metropolitan counties which would correspond more neatly with the reality of late twentieth-century Britain. For some people, these entities were difficult to identify with in any meaningful way. Older deep-rooted senses of attachment to county proved deeply ingrained and persisted despite the formal disappearance of some counties. Significantly, 'old' counties were re-instated in the 1990s, in part perhaps in recognition of people's attachment to older territorial divisions and the failure of newer ones to gain broader acceptance in the public mind. Old county divisions never fully disappeared and county loyalty was affirmed and perpetuated through a variety of practices, through county cricket teams, local newspapers and other media which served to reinforce people's self-identification. These continued to operate and helped to

maintain (and perhaps even strengthen) people's attachment to a particular construction of place and to reinforce place boundaries. This adds
further strength to the arguments in earlier chapters concerning both a
sense of place and the extent to which boundaries, once created, become
powerful elements in shaping identity. It can be concluded that administrative divisions may well have considerably more meaning in our
everyday lives than simply lines on a map or providing a mechanism for
determining service delivery. This is a further example of the ways in
which the state impinges on everyday existence and of how a sense of
territorial identity may well reinforce state structures. Of course, it is also
the case that a sense of identification with these territorial divisions can
be a useful mobilizing force for activities which can be seen as subversive.
Various protest groups may organize themselves into county-based units
to protest against what they see as undesirable activities on the part of
the state such as hospital closures or cutbacks in transport services (Figure
7.4). Once again, territory is being used as an organizing device in order
to attain political ends.

A good example of where primarily administrative divisions have
successfully engendered a sense of identity is the Irish county system.
In Ireland, county loyalty is important for many and nowhere is this more

Figure 7.4 Demonstration against government spending cuts, Worcester, England
(Source: Author)

apparent than in the playing of Gaelic games. The Gaelic Athletic Association (GAA) is the body which administers the sports (most notably hurling and Gaelic football). The GAA's organizational structure has a strong geographical base using Ireland's provinces, counties and parishes as its key units. This has helped to inculcate or reinforce a strong sense of place and an attachment to locality among its members where inter-parochial and inter-county rivalries are an important backdrop to fixtures. The association's major competitions are the annual inter-county All-Ireland football and hurling championships. Furthermore, this structure and the amateur status of the games means the organization is firmly embedded at a local level and teams remain strongly connected to local communities (Cronin *et al.* 2009). Another rather different way in which regions and localities are defined may relate to the production of specific products while linking to ideas of place identity, a good example being the regionalization of the French wine industry (Box 7.3).

Box 7.3 *Terroir*

There is an increasing trend towards highlighting the connections between foodstuffs and the area in which they are grown. There are a variety of reasons for this trend associated with such things as distinctive tastes, presumptions of quality, healthy lifestyles, environmental concerns, re-connecting producers and consumers and so on. These moves are further reflected in legislation protecting certain types of regional food production. The EU-wide system of Protected Designation of Origin (PDO) and Protected Geographical Indication (PGI) serves to restrict the use of local food names to those produced within the region of origin; an example is parma ham which can only be called this if produced in the region around Parma in Italy. A long-standing version of this type of linkage between product, process and place (pre-dating EU legislation) is evident in the wine industry. Wine production in many countries emphasizes the importance of the area of origin to the final product. In France, the term '*terroir*' encapsulates the importance of place to wine production. The term does not easily translate into English referring to the very localized assemblage of soil, climate, topography, aspect and culture, a combination that is presumed to impart a specific character to a place and the wine produced there. *Terroir* is reflected at a variety of spatial scales and at a national level through the distinctive wine-producing regions of France such as

Bordeaux and Burgundy. Within each region, the best wines are indicated by an *Appellation d'origine contrôllée* (AOC) designation. This is a strict quality system dating from the 1930s, but constantly evolving, involving precise specification of geographic delimitation, grape types, growing and production techniques, and production maxima. Each AOC has acquired its own reputation and at highly localized levels the names of individual communes are inextricably linked to the wine produced there. One example is the wines from Chateauneuf-du-Pape. This area produces red wines using Grenache and Syrah (Shiraz) grapes in combination with other varieties (Figure 7.5). The AOC system serves to link production to the social, cultural and environmental characteristics of specific places. It also engenders a sense of place-based advantage for some growers in which there is strict regulation of the grape types used and the growing practices employed. Such systems offer a form of territorial protectionism to producers of place-specific products and these local products can also act as the peg on which to hang place promotional strategies and to develop gastronomic tourism (see Unwin 1991; Moran 1993; Barham 2003; Ilbery *et al.* 2005; Maye *et al.* 2007).

Figure 7.5 Vines at Chateauneuf-du-Pape, France (Source: Author)

Community and locality

Another dimension to this governance debate and the shifting role of the state is the current emphasis on voluntary and community groups engaged in local development in both urban and rural contexts. Traditionally local and regional development has utilized a 'top-down' paradigm with solutions and programmes implemented by professionals in the planning and development arena. This 'top-down' approach was much criticized because of its neglect of the views of local residents resulting in the advocacy of more 'bottom-up' approaches centred on the inclusion of the concerns of local people and the empowerment of those people so that they actively participate in devising strategies for their own areas. An additional argument is the perceived need for locally specific responses to those global processes which have tended to open up localities to a range of external forces (Woods 2007). Consequently, the contemporary development discourse emphasizes the importance of participation by local agents alongside an espousal of the centrality of community involvement, empowerment and capacity-building. It is argued that the involvement of local individuals and agents utilizes local knowledge and skills and it may ensure more success given that it enjoys a degree of local support while tapping into people's identification with locality and community (Moseley 2002). It is therefore suggested that this participatory approach is not just more democratic, but also more effective.

This territorial framework may also serve as a mechanism which induces a real sense of community, thereby mobilizing local residents and facilitating positive practical outcomes. This operationalizing of regional identities is seen as important in local development, and community is seen as a potentially positive resource, particularly in a rural context (Ray 1998; Roca and Oliveira-Roca 2007; Oliveira *et al.* 2010). A further refinement of this process is the promotion of ideas of partnership between statutory agencies, the private and voluntary sectors. Partnerships come in many different forms, they operate at (and across) a variety of spatial scales and they may be a source of genuine added value, not just mechanisms for collaboration (Edwards *et al.* 2000; Moseley 2003). Territorially defined communities are envisaged as playing an integral part in the process of initiating and managing projects in their own areas.

Viewed in a wider context, these shifting organizational arrangements reflect changing modes of governance where power and authority is assumed to extend beyond the formal structures of central and local government to embrace the interactions between a diverse range of actors at various levels in the spatial hierarchy and having varying degrees of influence. There is a shift away from centralized forms of state regulation to what might be interpreted as more flexible modes of local regulation. We seem to be witnessing

> new styles of governing that operate not only through the apparatuses of the sovereign state but also through a range of interconnecting institutions, agencies, partnerships and initiatives in which the boundaries between the public, private and voluntary sectors become blurred. The actors and organizations engaged in governance exhibit differing degrees of stability and longevity, take a variety of forms and operate at a range of scales, above, below and coincident with that of the nation-state.
>
> (Woods 2005: 164)

These moves to a more participative and partnership-based approach might be seen in a positive light seeming to encourage decentralized decision-making, recognizing the role of local people in the development process, fostering indigenous potential and breaking down divisions between 'professional' and 'amateur', between local resident and external 'expert' while emphasizing the importance of diversity and local difference. However, leaving aside technical issues over the operationalization of partnerships and how best to evaluate the effectiveness of programmes, more critical attention needs to be paid to issues of the nature of community and power.

One key problem here is the use of the term 'community' to imply a harmonious and inclusive emphasis on the local. Community is a term with potent meaning and one often 'swathed in soothing feelings' (Staeheli 2008: 7). Despite its appeal, some simplistic interpretations conflate communities of place with communities of interest where a commonality is assumed between all those resident in a geographically defined area. Territorially based partnership approaches to development tend to elevate residential location and the 'local' while brushing over various fault lines such as class, gender, or length of residency in the area (Liepins 2000). Underlying much of the rhetoric appears to be an assumption that partnership and participation are intrinsically good things and the supposed merits of the approach have become acts of faith not to be questioned (Cleaver 2001). In emphasizing community and encouraging the construction of partnerships, developmental initiatives endeavour to find common goals whereby potentially divisive issues may be sidelined in favour of harmony, resulting in what some suggest is a standard set of interventions (Mosse 2001). This may mean that 'easier', achievable and less contentious options may be pursued, while more problematic issues are ignored in order to minimize potential friction.

Despite the rhetoric, it is not necessarily the case that partnership arrangements equate to widespread participation. Some voices are more likely to be listened to while others are ignored or remain silent. It may be the 'usual suspects' (those with the time and the energy) who are heard regularly while a 'silent majority' sits outside the consultative structures (Lowndes and Sullivan 2004; Osborne *et al.* 2004). There are additional risks that individuals and groups become incorporated into the development structures, allowing more powerful partners, whether central government, local authorities or other statutory bodies, to 'capture' the agenda and to sideline dissent. In effect, the discourse of participation, community, empowerment and so forth disguises the various power relations in operation whereby local people are 'involved' but are not necessarily in control.

Alongside questions of who constitutes the local community (who is 'in' and who is 'out') referred to earlier, obvious questions arise here as to what the local actually is in terms of its spatial extent, its territory, its boundaries and its scale. There appears to be an implicit idea that the local can be both unambiguously defined and lauded as unequivocally 'better' (Harvey 1996). An emphasis on the local, while it may have potentially useful outcomes, may also create or magnify competition and divisiveness between localities as each tries to out-do the other in drawing attention to their 'deserving' status. This can have the effect of diverting energies into funding bids and creating a case for funding rather than into what might be the more productive activity of devising development strategies. While it may be sensible to query the supposed external 'experts', it does not necessarily follow that local solutions are intrinsically better merely by being 'local'. It can reasonably be argued that local knowledge, no more than expert knowledge, does not itself possess an intrinsic value. It is far from certain that a knee-jerk lauding of the local, in so far as one distinct local can be identified, is in any way intrinsically 'better' that an externally imposed view as local knowledge is not fixed, it is contingent and is shaped by various interests (Mosse 2001). Consequently, there may be a danger 'of swinging from one untenable position (we know best) to an equally untenable and damaging one (they know best)' (Cleaver 2001: 47). The replacement of a 'top-down' orthodoxy by another participative one to which all must subscribe is seen by some as merely a new form of tyranny rather than heralding a new and radical approach to development (Cooke and Kothari 2001). In addition, the more romanticized justifications for local involvement tend to gloss over the possibility that some local perspectives may reflect insular and conservative values that can translate into exclusionary practices exerted against specific groups (such as gypsies) or individuals within a locality

(Harvey 2000). An emphasis on place-based communities may serve to legitimize forms of social exclusion.

This emphasis on voluntary activity at the level of local communities has also been in evidence in activities such as crime prevention, where people's sense of identification with their own neighbourhood is the key agent. Schemes such as Neighbourhood Watch in the UK are based on volunteers 'policing' their own localities and generally 'keeping an eye out' for anything unusual. The use of signs, displayed on lampposts, in windows and elsewhere, defines the boundaries of each Neighbourhood Watch area and acts as a clear territorial marker for those entering these zones (Figure 7.6). As with the other examples used above, it might be

Figure 7.6 Neighbourhood Watch sign (Source: Unisouth from Wikimedia Commons)

argued that this represents a cheap form of policing. It also ensures that local community activists remain 'onside' and become part of the policing apparatus rather than remaining outside it (Yarwood and Edwards 1995).

Summary

This chapter has introduced just some of the more obvious forms of sub-state territorialization. It has provided an analysis of the main forms of sub-state governance as well as introducing key examples of quasi-state territorial strategies. It has also applied the theoretical insights introduced earlier in the book to an analysis of this territorialization and has suggested that the interaction between the central and local states is far from simple as modes of governance become increasingly complex. What is very apparent, however, is the importance of territorial arrangements in the day-to-day running of the state. Territorial strategies are used in order to manage the state and to assure its reproduction. While these strategies can be seen in purely functionalist terms, it is important to bear in mind that the divisions used may help to foster or to reinforce a sense of identity – they are not merely functional containers. The encouragement of local and voluntary groups to engage in developmental activities and to become more actively involved in such arenas as local service provision provides a further dimension to the territorialization of everyday life.

Further reading

The works indicated here deal with some of the issues raised in this chapter including sub-state territorializations, community development and partnership arrangements, geographies of policing, and the connections between food and place.

Cooke, B. and Kothari, U. (eds) (2001) *Participation: The New Tyranny?*, London: Zed Books.
Cullingworth, B. and Nadin,V. (2006) *Town and Country Planning in the UK* (14th edn), London: Routledge.
Edwards, B., Goodwin, M., Pemberton, S. and Woods, M. (2000) *Partnership Working in Rural Regeneration*, Bristol: Policy Press.
Maye, D., Holloway, L. and Kneafsey, M. (eds) (2007) *Alternative Food Geographies.Representation and Practice*, Oxford: Elsevier.
Staeheli, L. (2008) 'Citizenship and the problem of community', *Political Geography* 27 (1): 5–21.
Yarwood, R. (2007) 'The geographies of policing', *Progress in Human Geography* 31 (4): 447–465.

8 Territory and society

As suggested in Chapter 1, territoriality is a phenomenon not confined to the world of formalized politics. Territorial configurations exist at a variety of spatial scales and are constantly reproduced, contested and transformed. At more micro-levels, territorial strategies may be used in attempts either to attain or to retain power or to achieve particular outcomes. This chapter focuses on issues related to social inequalities and indicates how, first, these inequalities are mapped onto space and, second, how social life affects, and is affected by, territorial formations. In the main, these are territorial phenomena which are less obvious, with less clear-cut boundaries, and which may be more difficult to detect and observe, relative to those discussed in earlier chapters. Although they may appear more nebulous, this does not make them any less 'real', as these 'informal' territories can still convey quite clear meanings to those concerned and are bound up with strategies of enforcement and control (Delaney 2005). While the processes occurring may be different, they can still be seen as spatial reflections of power and influence, and each of the examples chosen reflects the interconnections between geographical space and everyday life.

Thus far in this book, discussions of territories and territoriality have centred on what can be seen as the public arena of politics in its various forms. Feminist geographers in particular have been instrumental in focusing critical attention on divisions between the public and private domain and its spatial corollary of a division between what is seen as public space and private space. Within this chapter, reference is made both to the public domain, mainly divisions within urban space, and to the private domain with a focus on issues of class, ethnicity, gender and sexuality.

Underpinning the examples used here, in part at least, is a sense in which particular spaces become characterized as the preserve of a specific set of individuals. Boundaries are constructed and contested in people's everyday lives. These boundaries reflect broader issues related to power relationships: relations based on class, ethnicity, gender or sexuality which give rise to significant social inequalities. The construction of these boundaries engenders a sense of people being in their 'proper' place or 'out of place'. While many people do not necessarily freely choose their 'place', they may, nevertheless, identify with their immediate neighbourhood or locality. This sense of identity can in turn be converted into forms of action aimed at obtaining particular outcomes. The formation of community or residence groups reflects feelings of belonging or attachment to a particular place. In line with earlier discussions, it follows that notions of territory are connected with ideas of social power. The claiming of space is a political act whether it occurs in the public or private arena. In line with earlier discussions, the object is to examine the ways in which social phenomena manifest themselves territorially and how territorially based strategies may be utilized in the pursuit of certain objectives.

The examples are organized into five main areas: class, 'race' and ethnicity, gender, sexuality, and work, rest and play. These provide a framework for the exploration of forms of territorialization and a consideration of geographies of resistance. Despite the adoption of this structure, it should be abundantly clear that many of the issues raised are interrelated. People have more than one single identity: gender, sexuality and ethnicity cross-cut each other in a system of overlapping identities. Ethnic groups and classes are not immutable but are social constructs. In other words, they vary in time and space and are not simply natural categories, but human (that is social) products. Rather than seeing identities as rigidly defined, it is more useful and realistic to regard them as unstable, relational and contingent. The examples used in this chapter raise obvious questions about identity, rights and belonging thus touching on broader ideas of citizenship (alluded to in Chapter 3). While traditional notions of citizenship are centred on the relationship between the individual and the state, with a focus on rights and duties, more recently, ideas of citizenship have broadened to consider group rights and issues below the level of the state (local issues) and beyond the state (responsibility to distant others). While people may appear to enjoy full citizenship rights, as we shall see, particular territorializations and the use of territorial strategies can result in some groups being excluded from participating fully within society. In effect, some people may be made to feel like second-class citizens as a consequence of the creation and maintenance of informal territorial boundaries. This is one reason why various social movements and groupings have arisen to assert group identities and rights often through challenging particular territorializations. Consequently, this chapter is concerned with raising questions about who is 'allowed' to be in particular spaces and who is barred or discouraged from being there. In doing so, it sheds light on the ways in which space is conceived, used and organized at localized levels (Elden 2005).

Class and segregation

Society is seen to be stratified into different socio-economic groupings. An analysis of these social classes casts light on inequalities between different sections of society: between rich and poor; between property owners and those who are not; between those who own and control resources and those who are paid workers or are unable to obtain a job. These social divisions (together with race, ethnicity, gender and others) not only reflect inequality, but are also deeply spatialized. For example, most cities appear to have distinct residential neighbourhoods, collo-quially defined as 'rich' or 'poor', 'working class' or 'middle class'. These spatial divisions are reflected in ideas of 'the wrong side of the tracks' and everyday discourses of 'good' and 'bad' areas within cities. These latter zones are often viewed as separate spaces inhabited by many people who are marginalized not just in social and economic terms, but also spatially. Sidaway (2007) has used the idea of enclaved space as a way of thinking about social (and other) divisions which serve both to reflect and to reproduce inequality. One good example is the *banlieues* of Paris which are both physically separate and socially distant from the heart of the city, cut off by a ring road (the *périphérique*) from the centre and relatively isolated from each other (Morley 2000). Dikeç (2007) high-lights the ways in which the *banlieues* have become synonymous with high unemployment, crime and related negative images. Their residents suffer discrimination and inequality, while being subjected to negative stereotyping and harsh policing. They have come to be characterized as dangerous neighbourhoods inhabited by dangerous people patholog-ized as an underclass. Urban riots in the poorer *banlieues* on the edge of Paris in 2005 (and smaller scale events in 2007 and 2009) spread to other French cities. Young people there, including many from families of North African migrants, reacted to incidents of police violence in a wave of protests that rapidly spread. In these areas of high unemployment, younger residents gave violent expression to their feelings of social and spatial alienation. The urban unrest has deep structural roots embedded in the French urban planning system and in wider French political struc-tures (Dikeç 2006). Such stigmatizing of place (and people) in itself becomes part of the problem, serving to reinforce class (and often ethnic) divisions and to reproduce various forms of social exclusion. This can result in ghettoization with a whole series of negative consequences for both people and place so that the 'inhabitants come to wear the myths of that place' (Winchester *et al.* 2003: 176).

In general terms, these and similar forms of segregation arise and are reinforced through various mechanisms, particularly the housing market, which effectively determines who can afford to live where. The 'poor' cannot afford to live in 'rich' areas and the rich are generally unlikely to want to reside in close proximity to poorer run-down neighbourhoods. Nevertheless, attempts to regenerate older working-class areas tend simply to re-shape the geographies of class rather than eliminate them. Gentrification is where parts of the urban area experience regeneration or renewal resulting in more affluent residents moving in and displacing the original predominantly working-class inhabitants. Driven in part by economic considerations and in part through consumer choice, it serves to reinforce economic divisions within society thereby perpetuating the idea that some households do not belong in particular places (see Hubbard 2006). Linked to this, it is argued that the role of 'urban gatekeepers' (such as estate agents) may play a key role here in altering (or endeavouring to maintain) the social composition of particular areas (Shaw 2008). Gentrification reflects broader socio-economic processes and the resultant residential territorializations can be seen as an expression of both demand for housing (from home-owners) and supply of capital (from financial institutions). It serves to highlight how broader global processes play out alongside (and are implicated in) the destruction and reconstruction of local territorialized identities (Butler 2007). The major regeneration schemes undertaken in recent decades in older industrialized and dockland areas in cities throughout Europe and North America reflect this transformation from manufacturing and working-class residential spaces into service sector zones with middle-class residents. Dockland and waterfront areas, like other 'regenerated' urban zones, have been transformed into different places, with quite different uses and symbolic meanings. In South Wales, for example, the regenerated Cardiff Bay area has been transformed from a place associated initially with the export of coal and subsequently with industrial decline and decay (Figure 8.1). More recently, with its mix of residential and commercial developments, it is promoted as a symbol of a modern forward-looking city. The scale and extent of contemporary developments, in global cities like New York and London, where a super-rich elite engages in intense investment and conspicuous consumption, has led to the coining of the term supergentrification (Butler and Lees 2006). These attempts to purify urban space have led to the displacement of some urban residents while others are rendered homeless. In cities such as San Francisco, fragile lives are lived out on the streets amidst an atmosphere in which the homeless are

Figure 8.1 Cardiff Bay (Source: Author)

seen as a blot to be removed rather than as a manifestation of a systemic problem. Homeless people are criminalized and medicalized so that urban space can be cleansed and put to more profitable uses (Gowan 2010).

Class-based segregation is rendered even more obvious through the long evident phenomenon of urban gating (Davis 1990; Blakely and Snyder 1997; Atkinson and Blandy 2006; Glasze *et al.* 2006; Bagaeen and Uduku 2010). The apparent rise in the numbers and varied forms of gated communities within urban areas in various parts of the world in recent years could be interpreted as a very obvious manifestation of attempts to control and limit access to portions of geographic space (Figure 8.2). The creation of residential fortresses where security guards patrol the perimeter of walled residential zones in an effort to exclude those seen as 'undesirable' has been repeated and deepened in many other cities in order to exclude those seen as not belonging there, so maintaining the 'undefiled' and 'exclusive' nature of the neighbourhood. The level of 'fortification' of these developments is quite varied ranging from perimeter walls, gates and barriers, through the limiting of non-residential access by intercoms and associated 'screening' devices, to more perceptual barriers or codes deterring access.

Urban gating has a long history and has evolved in different forms in different parts of the world (Le Goix and Webster 2008). Many contemporary versions of gating can be seen as a consequence of two interconnected factors: security and prestige. Discourses of safety and security serve as a useful rationale for developers to design, build and promote spatially exclusive housing often surrounded by security fences and with

Figure 8.2 Gated residential canal-side housing development, Worcester, England
(Source: Author)

highly limited public access regulated by intercoms and other 'screening' technologies which have come to be regular features in contemporary landscapes of power. These territorial strategies work in ways which ensure a particular residential mix and may well serve to link together both racial and class divisions. Ideas of exclusivity and a quality lifestyle in a 'dream home' feature prominently in the way in which such developments are marketed. Gating is not strictly the preserve of the better-off and forms of gating take place in cities such as Accra and in the slum areas of cities such as Mumbai as well as in the gentrified zones of European and North America cities (Grant 2005; Nijman 2009). More broadly, Morley (2000) suggests social groups separate out from each other and, indeed, are encouraged to do so. He further suggests that these tendencies have been effectively reinforced in recent decades through geo-demographics and the use of postcode data and associated marketing strategies of companies who are keen to identify particular types of consumer, and link these to geographic areas. In this way, residential homogeneity is both reflected and reproduced. The social and the spatial are inextricably linked as 'gated minds' are translated into gated places (Landman 2010). While the forms it takes may vary somewhat according to place-specific circumstances, gating in its various guises touches on vitally important questions of the privatization of previously public space, inclusion, exclusion, and the territorialization of social life (Lemanski and Oldfield 2009; Rosen and Razin 2009).

The mixture of subtle and not-so-subtle processes of residential segregation is mirrored in other arenas. Taken alongside the privatization and 'gating' of residential zones, the proliferation of covered shopping malls can be seen as the erosion of shared urban street space and its replacement with privatized, more exclusionary spaces of consumption. Where once streets were open to a broad public, there are now privatized spaces whose owners (invariably resorting to security firms who take on some police powers) can evict those seen to behave inappropriately or who simply look 'out of place'. The shopping mall has become a privatized and highly regulated space in which people may be excluded by virtue of their appearance or behaviour (Staeheli and Mitchell 2006). In the United Kingdom, recurring debates and associated social panics over young people wearing 'hoodies' (hooded sweatshirts popular among teenagers) reflects particular hegemonic ideas about how teenagers and young adults should behave in socially acceptable ways in particular places. In 2005, a shopping mall in Kent, England, banned young people from wearing 'hoodies' and baseball caps as part of a crackdown on what was deemed to be antisocial behaviour. The centre is considered private space into which people are 'invited' rather than having any automatic right to be there. These 'secure' shopping centres, office blocks and apartment buildings, complete with gates and intercom systems, exemplify a trend towards socio-spatial design whereby territorial strategies associated with crime prevention effectively exclude those not wanted. Some people are effectively barred from certain areas – a policy of territorial containment enforced through increased surveillance and architectural design features. In part, these forms of exclusionary policing reflect broader geographies of fear and the perceived risk of crime associated both with specific groups and with specific geographic spaces (Pain and Smith 2008).

Another form of spatial enclaving is the proliferation of exclusive recreational spaces such as golf courses, wider tourist complexes in regions such as the West Indies, or beach resorts in parts of south-east Asia and elsewhere. These are designed to enclose the relatively privileged consumers from the excluded local 'other' (see Bunnell *et al*. 2006). Viewed in a broader context, the production of these spaces of exclusion can be interpreted as a long-standing and inherent tendency within capitalist social relations bound up with constant pressures towards the privatization of the public domain (Harvey 2003; Vasudevan *et al*. 2008).

Ethnicity, 'race' and space

Just as class is mapped on to the urban fabric, so too is ethnicity. While ideas of race appear firmly embedded in everyday discourses, racial and ethnic categories are social constructions rather than simple biological realities. Although race can be questioned as a problematic and often dubious form of social classification derived from past notions of hierarchy and domination (notably slavery and colonialism), there is no doubt that racism or 'race thinking' is a very real social phenomenon (Saldanha 2011). Though the idea of a 'scientific' categorization of humanity into different 'races' has long ago been challenged and is now thoroughly discredited, its legacy persists. Kobayashi and Peake argue that race is socially constructed but that 'racialization' is 'the process by which racialized groups are identified, given stereotypical characteristics, and coerced into specific living conditions, often involving social/spatial segregation' (2000: 293) thereby producing racialized places. While issues linked to 'race' are clearly social phenomena, they are often manifested spatially. Among the most rigid examples of racialized space was that devised under the apartheid system in South Africa from the late 1940s to the early 1990s: a territorial system that enhanced and entrenched the political, economic and social power of a minority white population over non-white populations, as discussed earlier. Both nationally and at the more localized level of individual urban areas, space was divided on racial lines. Non-white people were 'placed' in locations not of their own choosing in order to entrench minority white power. In this way, there was a legal transposition of inequality on to geographical space. This spatial arrangement was designed to ensure greater degrees of control over the majority black population and is a classic example of the utilization of a territorial strategy to attain political objectives. At very localized levels, there was a racialization of space with buses, public toilets and other amenities reflecting this divide. A racial ideology was mapped on to the South African landscape. Although apartheid ended in the early 1990s, after a long struggle for democratic rights for all in South Africa, its legacy means that a division of space based on the racial and class lines reinforced during the apartheid era has left enduring marks on the social landscape.

Elsewhere, nationalist conflicts may give rise to attempts to create ethnically pure spaces through the forcible removal of ethnic 'others'. The phenomenon of ethnic cleansing in parts of the former Yugoslavia during the 1990s is a striking example of an attempt to 'purify' territory of those deemed to belong to other ethno-cultural groups in an apparent attempt to justify territorial control in the name of the nation (see Chapter 5). In a not unrelated manner, the generally negative stereotyping of gypsies in much of Europe has led to considerable discrimination, with gypsies seen as an undesirable 'other', as a consequence of which they are effectively de-territorialized: they are seen not to belong anywhere and active attempts are made to exclude them from certain spaces (Fonseca 1996). In 2010, the French government attempted to repatriate ethnic *roma* people to Romania, in what appeared a contravention of EU policy on freedom of movement between member states, as well as a highly questionable attack on an ethnic group. Broader issues emerge here in relation to groups which pursue nomadic lifestyles and who, consequently, are subject to considerable public opprobrium. Attempts by gypsy and traveller groups to use particular spaces are often resisted by settled residents. As a group, they are de-territorialized, simultaneously belonging everywhere and nowhere, their mobilities juxtaposed to the settled nature of place-based communities. The mobile lifestyles of some are seen as unnatural and those who practice them are often depicted as untrustworthy and are subject to a range of discriminatory practices (Shubin 2011).

Leaving aside these overt and coercive examples, it is clear that many cities exhibit high degrees of ethnic segregation. In most US cities, for example, the elites and middle class are disproportionately white. The spatialization of class (discussed above) thereby contributes to ethnic segregation (Crump 2004). These patterns of exclusion and inclusion and attendant territorialities reflect the complex intersections of race, class and ideology. In considering the evident spatial concentrations of ethnic groups in urban areas, it might be argued that individuals choose to locate in such areas for a variety of reasons. In brief, there are a combination of 'positive' and 'negative' factors; for some there are attractions such as 'being amongst one's own', while others may feel driven to seek sanctuary from a racist, hostile society. A distinction might be made between a structurally influenced segregation and self-segregation. Following Knox and Pinch (2000), key reasons may be summarized as follows. Clustering affords *defence* against attack by the majority group. It also provides a degree of *mutual support* and bolsters a sense of belonging and community. This in turn may be a useful means through which group cultural norms and heritage may be preserved. Finally, clustering produces *spaces of resistance* whereby external threats, whether to cultural norms or of physical attack, may be reduced. However, these arguments should not detract from the fact that residential clustering may be more a function of necessity rather than free and unconstrained choice. The degree and extent of choice available to many may be extremely limited. The idea that people may choose to cluster is to ignore the fact that quite often no easy alternatives are available. Discriminatory ideologies of race work to exclude people from particular areas, thereby translating social exclusion into geographical exclusion. 'Ghetto' areas are institutionally produced and reflect individual and institutional racism. They are far from being simply the product of choice by those who come to inhabit them (Crump 2004).

It is also clear that such clustering itself further contributes to future rounds of marginalization and exclusion so that the spatial divisions resulting from economic and social marginality have a tendency to lock people into a socio-spatial milieu which impedes economic advance or social mobility (Smith 2005). This further demonstrates the idea that territorial practices serve to reproduce particular social outcomes in a complex relationship whereby society does not simply impact on space, but spatial arrangements in turn impact on society. In considering so-called ghetto areas, there is a tendency to see them in quite negative terms. The ghetto is a territorial entity and it is one that evokes many negative connotations: the term is often seen as synonymous with 'slum', the juxtaposition leading to a stigmatizing of its residents. In such ways, the hegemony of the dominant group is maintained and the 'other' (in this case, ethnic minorities) remains geographically, as well as socially and economically, marginalized. In Western cities, 'ghettos' and other 'inner city' areas are frequently seen as the home of a so-called 'underclass' experiencing relative and absolute deprivation and disadvantage. Places such as South Central Los Angeles or the *banlieues* of Paris are represented as dangerous areas to be avoided. In this way, the 'ghetto' is portrayed *as a problem* rather than a place whose residents *experience* problems. It is also depicted in a somewhat monolithic way that ignores its internal social, economic and cultural diversity. Before proceeding to look at further micro-scale territorializations, it is useful to highlight the ways in which spatial concentration and segregation may be used as a means of resistance by subordinate groups.

Geographies of resistance

The sorts of territories discussed above can be seen as manifestations of social inequality reflecting the uneven distribution of power in society. While there are many negative consequences of territorialization built on ideas of race, class or ethnicity it is dangerous to paint places as poor marginalized ghettos whose residents lie somehow passive and unable to resist the forces seemingly arrayed against them. Slum areas in cities are people's homes, the places in which their daily lives are lived out. In cities such as Mumbai, as many as half of the city's population live in areas that could be categorized as slums. Despite what may appear to the outsider as random collections of buildings and activities, areas such as Dharavi in Mumbai exhibit patterns of spatial organization reflected in internal territorializations in which specific functions (such as industry) take place in particular zones. These areas provide people with senses of identity, belonging and community, and may be perceived as safe spaces from the world beyond. Despite their negative images, these are places where economic, social, cultural and political activities take place and in which social networks bind people to place (Nijman 2009).

The apparent consignment of the poor or of ethnic groups to certain areas can also of course provide the spatial framework for forms of resistance. The 'ghetto' can be used as a means of mobilizing residents, of providing the territorial frame with which people identify and within which they can operate with a view to improving their own conditions. Successful residents' groups often emerge and may become sufficiently well organized to be able to engage in lobbying, in forms of self-help, and so on. Such community action may well make a positive difference to the lives of ghetto residents. Nijman (2008) provides a good example of a poor community in Mumbai engaging in a housing improvement project with some degree of success. Such clustering may facilitate the election (at local level at least) of political representatives for residents of those areas thereby providing them with a voice that might otherwise be denied them. It may also facilitate the mobilization of people in support of, or in opposition to, issues of direct concern to them such as transport, banking facilities, the provision (or non-provision) of a range of services and the preservation of open spaces. In this way, a territorial strategy can be utilized in order to defend the interests of those who, if more spatially scattered, would be unable to do so. Community groups in working-class areas in European cities or in the *barrios* and *favelas* of many Latin American cities exemplify this tendency. Resistance may also take more overtly subversive forms such as street gangs laying claim to their 'turf', the renaming of streets and the placing of territorial markers (Box 8.1). Ultimately, the contestation of space through various forms reflects resistance and an assertion of rights and indicates the ways in which the meanings of place and community are disputed.

Box 8.1 Urban gangs

The activities of urban street gangs have received considerable media attention in recent years, particularly in the United States. These gangs engage in occasionally quite violent conflict and their behaviour has a strong territorial component. Their *raison d'être* is the assertion of control over their 'turf'. Rival gangs are not welcome on their patch. This territorial behaviour might be seen as reflecting some sort of innate territoriality, but it can also be interpreted as a consequence of the marginalization of many poorer young people in impoverished urban areas. Territoriality may be a means of expressing power using the main resource available to them, the streets and neighbourhoods in which they live. These gangs may be linked to criminal behaviour and may be involved in controlling illegal activities in their patch, mirroring the behaviour of 'older' criminal gangs who also display a territorially based organizational structure.

These gangs often lay down territorial markers to indicate to others their 'ownership' of particular places. Graffiti on walls, bridges and buildings is one very visual method of claiming space. Markers are quite literally placed on the landscape to signal control of territory or 'turf ownership'. Work in Philadelphia indicated that graffiti became denser closer to the core of that gang's territory. Ley and Cybriwsky (1974) were able to demarcate reasonably accurately the spatial extent of gang control in the city. In this way, aspects of popular culture are translated into a territorial frame. This claiming of space may be a means by which marginalized youth make their claim to existence: drawing becomes a territorial act (Brighenti 2010b). For young people living in gang areas, there are multiple problems of exclusion, marginalization and victimization resulting from both the activities of gangs and the ways in which such places are policed and stigmatized (Ralphs *et al.* 2009).

Places associated with visible ethnic minority groups can seek and develop positive connotations. Chinatowns in Western cities, for example, have often become sites on tourist and gastronomic circuits. London's Brick Lane, an area with a long history of immigration, has latterly come to be seen as epitomizing a certain cosmopolitanism, which can be mobilized as a resource (Dwyer 2005). However, while there may be many potentially positive outcomes to this territorial construction, it may also be contested and we need to be careful not to romanticize places through simplistic readings of ideas of 'local' and 'community' (Staeheli 2008; Storey 2010). Residents may not agree on the way in which the area is represented or portrayed. For example, there was a hostile reaction from some local residents to the fictional portrayal of the area by Monica Ali in her novel *Brick Lane* (2003). Another outcome has been the commercialization of the area and its 'culture', with tourist guidebooks promoting its restaurants. The area also hosts festivals and music events celebrating cultural diversity (see http://visitbricklane.com). While there are clearly positive dimensions to this, such strategies also run the risk of essentializing the identity of both places and their inhabitants. There is a risk of the portrayal of such 'ethnic spaces' as centres of exotica, to be exploited for commercial purposes thereby further contributing to the contesting of identity and meaning in such spaces.

We should also be mindful of the fact that much discussion surrounding issues of 'race' and ethnicity in Western societies tends to assume 'whiteness' as the norm. As McGuinness (2000) argues, much progressive research itself falls into this trap with a focus on non-white groups, tending to deflect attention away from white ethnicity. One consequence of the pursuit of this 'new exoticism', as McGuinness terms it, is that relatively little attention is given to 'white spaces'. Ideas of white flight to the suburbs (in response to the evolution of 'black' ghettos) and the creation of 'white' territories are themselves elements in the racialization of space. Similarly, the construction of rural Britain as a relatively 'white space' reflects deeply embedded ideas associated with belonging, rurality and with national identity (Holloway 2005; *Journal of Rural Studies* 2009). Such constructions can have serious implications for those who do not (or are seen not to) 'fit in' with the dominant assumptions and ethos. Recent debates over immigration into the UK and other western European countries have tended to focus on a supposed 'invasion' of asylum seekers, 'illegals' and 'hordes' of eastern Europeans (Gilmartin 2008; Samers 2010). While much of this is inaccurate and misleading, the nature of the comments suggests an idea of the United Kingdom as a territorial entity that should increasingly seal itself off from invasions from 'outside' by those who do not 'belong' here. These arguments are often (misleadingly) bound into security discourses emphasizing the need to protect the country and its citizens. Territory, terrorism and identity become inextricably linked in debates calling into question who has the right to be in certain places and who has not (Elden 2009). Additionally, countries are seen to extend their border controls well beyond their own territory with immigration personnel screening passports in foreign airports in order to interdict those deemed 'illegal' (Mountz 2009).

While racist and exclusionary ideologies are transposed on to space, this section has highlighted ways in which those social constructions are opposed. Just as particular power relations are refracted through a territorial frame, so those relations are contested through territorial strategies. The spaces to which people are consigned may provide the means through which they contest their marginalization. A territorial base may serve as a means through which an ethnic identity or a class identity is reinforced and reshaped, in part at least, in opposition to other identities. Other categorizations, such as religious affiliation, may also provide the basis for parallel territorialities. Of course, we need to be mindful that such identities may be deeply contested and are far from monolithic, though there may often be attempts by some 'inside' or 'outside' to portray them as such.

Gendered space

The growth of what became known as the 'feminist movement' from the 1960s onwards, building on earlier attempts to achieve equality for women, has been influential in gaining recognition in many societies for the unequal status of women and men in all dimensions of life; in the home, in (paid) workplaces, in the broader political and social arena. Feminist geographic interventions have challenged certain assumptions and have opened up new avenues of enquiry and new modes of analysis (Kofman 2008). Feminist writers and activists have been instrumental in attempting to explain how patriarchal systems of power have tended to reinforce male dominance and how women have often been marginalized (Dixon and Jones 2006; England 2006). Feminist geographers have drawn attention to the manner in which space and place are heavily gendered and have challenged the relationships between gender divisions and spatial divisions (McDowell 1999). Critical attention has been focused on divisions between public and private domain and its spatial corollary of a separation between what is seen as public space and private space. It is argued that patriarchal systems of power have led to a division between predominantly 'male' public and mainly 'female' private space, resulting in social practices whereby certain activities and certain spaces are seen as male preserves. This duality reflects broader distinctions centred on the binary divide between masculinity and femininity (Pratt 2005).

As with race and class, issues of gender are mapped onto space in various ways and the implications of gender are seen to be as important as other political, social and economic factors in the structuring of spaces and places. In its most simple form, this is reflected in the sexist notion that 'a woman's place is in the home'. The home has tended to be seen as a space of reproduction juxtaposed to the workplace as a space of production (Laurie *et al.* 1999). Underpinning this are ideas that distinguish between sex as a biological fact and sex as gender, which refers to the socially constructed roles of both male and female identities. In emphasizing the role of social conditioning, the argument is that as individuals we are not biologically predetermined to be more suited to some roles rather than to others.

One reason for the relative absence of women in particular places is overt discrimination or active discouragement in the sense of certain activities or pursuits not being deemed suitable for women. Historically, women who transgressed these boundaries were often portrayed in a negative light, an idea reflective of notions of 'good' and 'bad' women. Women out alone at night might be seen as not conforming to what is expected of them. A crucial aspect of the relationship between women and place centres on the perception of some specific places as 'unsafe'. Many women do not feel safe in certain public places, most notably darkened streets. As Valentine (1989) suggested, women transfer a fear of male violence into a fear of certain spaces, which has profound implications for the ways in which men and women negotiate their way through urban areas (Fell 1991). Clearly, the various strands of feminist thought and practice have resulted in significant advances with regard to equal rights for women. While this can be seen within the arena of equal pay and related issues, it is also reflected in terms of spaces. Thus, the heightened visibility of women in public space reflects the changing status of women. Phenomena such as 'reclaim the night' marches demonstrate the overt use of a spatial strategy to make a political and human point. While particular groups may find themselves excluded from certain spaces, those spaces can also be reclaimed (as an example see www.isis. aust.com/rtn/).

Historically, the gender division of labour tended to confine women to the private realm, leaving men to inhabit (much of) the public domain. This view of women as playing a subordinate role has in the past been reflected in discriminatory attitudes and practices, particularly in relation to women in the paid workforce, with active discouragement through lower wages, if not actual exclusion, from many jobs. These views are predicated on the undesirability of women going out to work. It can be argued that this ascribing of women's role, through delimiting the spaces in which women were encouraged to appear, is another spatial expression of power. In other words, the confining of women to domestic space, and their exclusion from male territories, was a key element in male control (Little 2002). With increasing female participation in the workforce and a raft of equal opportunities legislation in many countries, such a generalization may appear to have lost some of its validity. Nevertheless, the division between a (largely) male public sphere and a (largely) female private sphere still has considerable resonance in many societies (although the extent of this is itself immensely geographically variable across the world).

Where women enter the workforce, they may still encounter territorial divisions in the workplace. Thus, Spain (1992) documents the 'closed door' jobs of managers (mainly men) and the 'open floor' jobs of manual workers (who may be predominantly women in certain countries/regions/sectors). Employers may locate in particular localities (or countries) in order to take advantage of what they see as an available (and exploitable) workforce based on prevailing wage levels or skills and assumptions about gender roles (Hanson and Pratt 1995). The presence of women in the armed forces periodically provokes debate over the supposed appropriateness of women performing such roles. In 2007, Iran's arrest of 15 British sailors, one of whom was female, led to media commentary on the appropriateness of a woman, particularly a mother (as in this instance) being engaged in military activity in a war zone, rather than being at 'home' playing the key role in bringing up her children. The deaths of female soldiers in recent years in the wars in Afghanistan and Iraq have been depicted in especially poignant terms in the media. Social processes reproduce attitudes that tend to 'naturalize' a gendered division of labour in which women perform certain functions which are acted out in specific spaces; for example, 'home-making' and child-rearing in domestic space. Socially constructed gendered difference is inherently also spatialized. Within the arena of sport and leisure, gender stereotyping remains prevalent. While there are undoubtedly marked changes, gendered ideas and practices about leisure activities and, hence, separate spaces for men and women, are still common. Such social practices are built upon ideas of what is or is not acceptable behaviour for men and women to engage in (built on socially or culturally constructed notions of masculinity and femininity) and where stereotypes of women spending leisure time shopping while men attend sporting events, or watch them on television at home or in pubs, have a self-perpetuating quality. Media interest surrounding the shopping habits of the so-called 'WAGS' (wives and girlfriends of English international football players) provides further reinforcement of this.

If women have sometimes been seen (and in some contexts continue to be seen) as 'belonging' in or closely associated with the home, it is not the case that the space of the home is undifferentiated. Even within the home, territorial divisions take place, the most obvious being the notion of the kitchen as a female preserve. The relative neglect of such territorial behaviour in all its manifestations is commented on by Sibley, who suggests that interest in such things as residential patterns 'wanes at the garden gate, as if the private province of the home, as distinct from the larger public spaces constituting residential areas, were beyond the scope of a subject concerned with maps of places' (Sibley 1995: 92).

Many communities and societies often had a clear spatial division between men's and women's spaces and roles (Spain 1992). Within most Western societies, deep-seated ideas about woman's role as homemaker, cook, cleaner, child-rearer and so on mean that women have often been historically presumed to 'belong' in some rooms and spaces more than others. While such ideas have been challenged (and in many cases transformed), they sometimes endure. All these reflect territorial expressions of power whereby the designation or apportionment of space within the domestic sphere reflects the relative status or roles of the individuals concerned. Spain therefore reminds us that 'houses are shaped not just by materials and tools, but by ideas, values and norms' (1992: 111).

Sexuality and space

The idea of places territorialized by particular groups on the basis of sexual orientation has begun to receive more academic attention in recent years (see *Political Geography* 2006; Browne *et al.* 2007). At its most elementary level, this has seen the mapping of gay and lesbian 'zones' in selected cities. It is fair to point out that such spaces are not as easy to identify as, say, areas inhabited predominantly by a particular ethnic group. It is equally obvious that, in the main, these are not strictly demarcated areas. Rather, they are zones where gay people may feel more at ease through being accepted rather than rejected, scorned or ignored (or worse) by their neighbours. The mapping of 'gay territories' runs the risk of focusing on what some see as deviant behaviour as well as essentializing sexuality and reinforcing a gay/straight dichotomy. Nevertheless, the fact that those who identify themselves as gay and lesbian do, in some instances, become associated with particular (usually urban) spaces suggests that another form of territorial behaviour may be evident. A concentration of visibly gay restaurants, bars and clubs results in the creation of what Castells has called 'a space of freedom' (1997). Places such as the Castro District of San Francisco (Figure 8.3) and the more spatially confined 'gay village' in Manchester serve as important examples, while the city of Brighton in the south of England has acquired an image as the 'gay capital' of the UK (Browne and Lim 2010).

Figure 8.3 Castro district, San Francisco (Source: Matthew McPherson/Wikimedia Commons)

The construction of such zones may arise for reasons similar to those associated with ghettos and other forms of segregated space. Castells (1997) has argued that there are two key factors: protection and visibility. The first of these is fairly obvious. The idea of 'strength in numbers' may make people feel safer from homophobic 'gaybashers'. Visibility may have an emancipatory effect through which identification as gay or lesbian within a culture that is predominantly straight (heterosexual) and in which a straight discourse dominates may work to nullify views that see homosexuality as deviant or abnormal. Gay neighbourhoods become a means of asserting identities. Harry Britt, a one-time key figure among San Francisco's gay community, once commented that 'when gays are spatially scattered, they are not gay, because they are invisible' (cited in Castells 1997: 213). The significance of San Francisco's gay area was reflected in that community's ability to gain political representation. In obtaining power over territory, they also gained political representation and San Francisco has become recognized as something of a 'gay capital' of the United States with a somewhat more liberal attitude. In this way, the designation of 'gay territories' plays a crucial role in raising awareness of gay people and issues and also provides a means by which some degree of power and self-confidence can be attained. Celebratory events such as 'gay pride' marches can be seen as an assertion of citizenship rights through staking a claim to public space. Conversely, heightened visibility may render people targets for homophobic assault, both verbal and physical. The examples of gay and lesbian spaces suggest another important point – that of the temporality of territory. The longevity of these spaces may be quite brief as the 'scene' moves to somewhere else. Lesbian spaces may be very short-lived in time, whether caused by the transience of lesbian bars/clubs or the even more short-term phenomena of lesbian or gay evenings (Valentine 1995).

Of course, as with 'ethnic spaces' these territorializations may be inextricably bound up with factors that extend well beyond the realm of identity. 'Gay spaces' have often been associated with an economic imperative as the importance of the 'pink pound (dollar, or euro)' in aiding urban regeneration has frequently added a strong commercial angle to these developments. One consequence may be the 'co-option' of the identity in order to present an 'acceptable' image of the group concerned. Not surprisingly, some activists have expressed disquiet over the appropriation of such events and their dislocation from their original social and cultural roots and from their original territorial base. The fact that Manchester's 'gay village' and San Francisco's Castro District are firmly on the tourist trails of their respective cities may be lauded as an acceptance of identities previously scorned but it can also be seen as a commercialization of that identity which may, to some extent, serve to further ghettoize it (see www.sfguide.com/sights/neighborhoods/castro.htm). The complex intersections between different identities can create specific tensions which play out in different ways in different places. For example, the ways in which gay and lesbian identities are interconnected with processes of gentrification through which they take on a spatial dimension is apparent in places such as the Marais district in Paris where tensions between 'old' and 'new' residents and within the 'gay community' emerge (Sibalis 2004). Similarly, the intersections of sexualized and racialized identities in post-apartheid Cape Town in South Africa serve as another useful insight into how issues of identity play out in specific contexts (Tucker 2009). The relations between class, capital, sexual identity and place may be manifested through complex embodiments and manifestations of territoriality.

It might be argued that it is acceptable to be openly gay in an area so designated but in other spaces and societies the pressure to keep this identity hidden may well persist. While gay spaces may allow for more open expressions of sexual identity, homophobic assaults on sexual minorities reflect contested place meanings (Sumartojo 2004). Openly gay behaviour may be accepted or tolerated in some places but may remain decidedly unacceptable elsewhere. Here of course there are profound links between the small-scale (often neighbourhood or urban) territorialities and the wider policies of the city or state concerned. Many countries still criminalize gay sexuality and almost everywhere sexual mores and regulations are highly territorialized.

Work, rest and play

We can recognize two important social tendencies that bolster territoriality: the wish by people to have space of their own and the wish by others to exclude people from certain spaces. We have already seen ample evidence of the latter in this chapter. Even at what might appear to be very mundane or innocuous ways, the apparent claim to territory seems to manifest itself. At its most elementary level, the assertion of territoriality is reflected in claims to private property. Thus, people desire to mark their own home, to adorn it in their chosen style (influenced of course by social trends, technologies and fashions) and, in various ways, to mark it out as theirs. Home-owners are generally keen to stamp their personality on their home through the ways in which they choose to decorate it, alterations to layout, choice of colour schemes, furnishings and so on. This personalizing of space is further manifested through such things as the display of paintings, posters or photographs and the collection and arrangement of ornaments. The geographer Jean Gottman suggested that people 'always partitioned the space around them carefully to set themselves apart from their neighbours' (1973: 1). This manifestation is commonly interpreted as being symptomatic of our inherently territorial nature. This emphasis on the centrality of the home also has a broader cultural and political significance. Symbolic connections are often made between the domestic home and the nation whereby images of the former are seen to give material meaning to the latter. The home is seen in some ways to be at the heart of the nation. In times of war, for example, people have been encouraged to fight for the 'homeland' and the defence of 'hearth and home'.

Private property is regarded by many as an outcome of human territorial behaviour and it represents a claim to space that is reinforced by the legal system of many countries. However, as Alland points out, it might well be the case that 'private property is the child of culture and develops into a major preoccupation only with the evolution of complex society' (1972: 64). It follows that we need to be careful to avoid the trap of translating a need for personal space into an ideological claim for the sanctity of private property. The centrality of the family home, encapsulated in such phrases as 'home, sweet home', glosses over the fact that the privacy which many of us associate with the home is comparatively recent and is specific to some societies. Where 'domestic' space is limited (or for those who find it constrains them), life may be lived in the street (or perhaps the mall or car), much more evidently than the 'behind four walls' lifestyle many take as 'natural'. We need to be mindful of social, ethnic and geographic differences in the ways in which the home is conceived. In the United Kingdom, the notion of private home space was initially quite a middle-class idea which has since permeated the rest of society (Morley 2000).

Hegemonic ideas of the home within Western societies, exemplified by such notions as the home being an 'Englishman's castle' or associated with 'The American Dream', can be argued to have led to an ignoring of internal tensions and, in particular, a consideration of the different positions, roles and experiences of men, women and younger people within this domestic space, as discussed earlier (McDowell 1999). While the home is commonly depicted as a refuge from the outside world, it may also be a site for domestic violence and fear (Pain 1997; Squire and Gill 2011). Similarly teenagers may view the home as somewhere to escape from. In any event, the emphasis on the domestic idyll may have highly exclusionary consequences (Delaney 2005).

Even within buildings, territorial behaviour can be recognized. As we have seen, the idea of the kitchen as a 'woman's place' is one example of this; another is the notion of the garage or garden shed as predominantly a male domain. The domestic home in many different cultural contexts is often spatially divided, not just in terms of gender but also in terms of age, with certain spaces being designated for women or for children (Spain 1992). The banning of children from some rooms and the proprietorial attitude towards one's own room in a house are other examples of this. In the home, space is even being claimed at the level of 'my chair', 'my place at the table' and so on. There are also distinctions among those allowed in, with differential access for close family and friends on the one hand and more casual acquaintances on the other. Even then, friends may be welcomed into the living room but are less likely to be invited into the more 'private' spaces such as bedrooms (Morley 2000). In any consideration of the home, we need to be mindful that it is not a straightforward and unambiguous entity. While for some it conjures up feelings of comfort and security, for others it may be a place of discomfort, alienation and tension (Blunt and Dowling 2006). For some, the home may come to feel like a prison – quite literally so for those sentenced to home imprisonment of the type long imposed on political activist Aung San Suu Kyi in Burma or the many others subject to degrees of curfew, control and house arrest. Home is therefore a social construction in which social identities are (re)produced but conveying different meanings to different people.

Equally, in workplaces, some areas and rooms can only be entered by staff of a certain level and are out of bounds to more junior staff. These can be interpreted as managerial strategies designed to ensure a particular outcome: staff know their 'place' and can be more effectively controlled, sometimes through very obvious visual intrusion. Hanson and Pratt (1995) reveal how companies reproduce social segregation through spatial practices within the workplace whereby different sets of workers inhabit different parts of the factory and rarely, if ever, meet. Thus, office staff may be located downstairs in 'cubicles' separated by room dividers, with sales staff and management upstairs in individual or shared offices while production staff are located in an entirely separate part of the building. Socializing between workers tends to reflect their spatial segregation even to the extent of each department having separate annual parties. Work hierarchies are reflected in the spatial arrangements of the workplace. These practices have clear outcomes. They may render it difficult for workers to organize through physically keeping them separate and through engendering a sense of difference between different sections of the workforce. Distinctions between different grades of worker in the corporate hierarchy are matched to differences in the nature and quantity of allocated space. Ideas of more flexible practices may be reflected in 'hot desking' strategies whereby office workers are assumed to be able to function from any networked office space rather than necessarily from their 'own' space. While this might promote efficiency and cost savings, it does of course deny the worker the sense of a personalized space (Box 8.2).

Box 8.2 Playgrounds and lecture halls

There has been a growing interest recently in the geographies of younger people (see Holloway and Valentine 2000; Leyshon 2008), and as an example of micro-scale territoriality, it is instructive to consider the space of the school playground. Thomson (2005) suggests that while children's behaviour is clearly spatialized in a wide variety of ways, it might be thought that the playground is 'their' space in which school children can enjoy a degree of freedom to engage in a range of activities free from the constraints imposed in the more formalized and controlled spaces of the classroom or other parts of the school buildings and grounds. However, while play spaces can be seen as emancipatory, they are simultaneously regulating. Thomson's research indicates that on closer inspection the playground displays two key criteria associated with territories and territoriality. First, it remains heavily controlled by adults (teaching staff) who delimit the times it can be used and the activities which are permitted there. Second, it often contains internal territorial divisions in the sense of discrete spaces for different types of 'play' activity – sports, etc. What is also obvious, of course, is that children themselves will often endeavour to claim their own territory within the larger space of the playground with groups 'hanging out' in particular places. In addition, it is equally apparent that these spatial practices are open to contestation. Rival groups of children may 'compete' for spaces within the playground. What is also apparent is the manner in which children may try to test or push the spatial boundaries imposed by teachers and supervisors through such actions as encroaching on grass playing pitches and so on. Another important dimension of this (in addition to classification, communication, enforcement and resistance) is the way in which the control of territory is masked though recourse to other discourses such as that of health and safety – prohibiting access to spaces deemed dangerous or hazardous (Thomson 2005).

In a similar vein, we might take the example of the university teaching room. Here a lecturer may exert a strong degree of territorial control: occupying a space at the top of the room while controlling (or trying to) who else may speak and when they may do so. Even in highly interactive and more student-led sessions, a lecturer, by virtue of their status, continues to exert control over the space. She or he can 'command' the room and walk around it in ways which students cannot (or are discouraged from doing). However, such control has very obvious temporal constraints. What the students and lecturer may regard as 'their' room is usually limited to a regular timetabled slot. Before and after that time, their right to be there is denied as others lecturers and students take over the space and control it. Your geography lecturer is unlikely to walk into a lecture theatre during a microbiology class and commence to speak to the assembled students. (Well, they might try to do so but they would be met by a bewildered reaction and, more than likely, a phone call to security!)

Taking the idea of territory down to its most elementary level, the desire for personal space can be seen as a form of territorial behaviour. Humans like to have a pocket of space around them that is 'theirs' and will resent others 'invading' their space (unless invited!). This can be interpreted as a territorial claim to a portion of geographic space. While this might be taken as reflecting a natural tendency, it is worth noting that the amount of space needed appears to vary from one society to another, a fact noted long ago by Hall (1959). For many young people, their own room, apartment, etc. may seem like a 'natural' ambition but for many of the world's inhabitants, such a desire is completely unobtainable. For those living in overcrowded conditions, the amount of personal space available is extremely limited. For a homeless person, their 'own' space may be limited to a hostel bed in central London, a doorway in downtown Manhattan, or a small patch of pavement in the Tenderloin district in San Francisco. Nurture, culture, power and politics all need to be considered where the complexities and diversities of human territorialities are concerned.

Summary

The key argument of this chapter is that social practices are reflected in struggles and territorial claims over the use and control of space at a localized level. Territorial strategies are utilized in conflicts concerned with social power and identity at a micro-scale. This may be to do with maintaining power or with resisting the imposition of power by a dominant group. Forms of exclusion can be consolidated and reinforced through territorial practices, yet they can also be resisted through similar means. The examples provided are evidence of the ways in which social relations are expressed through spatial patterns and they highlight how these geographies help in turn to shape social relations. Social phenomena such as class, racial or gendered identities invariably embody a territorial component. Territorial strategies are often used to control and police those who are defined as 'out of (their) place'. In this way, particular ideologies are transposed onto space. People are confronted with wider practices through their use of space or through the ways in which they are allowed to use space. Power relationships take on a spatial dimension, even at the most mundane and everyday level. Issues of identity, particularly within multicultural societies, have a spatial expression as social divisions (associated with class, ethnic, religious, gender or other factors) are given material form through spatial divisions. The examples used here demonstrate the spatialization of wider ideas and they show how people are kept 'in their place' whether through overt mechanisms or more subtle means. Social boundaries are being communicated through space and 'the assignment of place within a socio-spatial structure indicates distinctive roles, capacities for action, and access to power within the social order' (Harvey 1990: 419). In this way, many territorial strategies are invariably discriminatory and exclusionary and can be used to deny people effective participation in 'society': the latter being based on a categorization of those who 'belong' (by virtue of citizenship and income levels that enable a range of choices and possibilities to participate).

However, as indicated earlier, just as dominant ideologies can be reinforced through territorial practices, they can also be resisted. Territorial strategies are useful mechanisms in the assertion of identity. Spatial concentrations within particular geographic areas make visible people and issues that might otherwise remain unseen. They can be used to draw attention to exclusionary practices and to assert the right to be equal citizens. In doing so, this demonstrates the 'positive' and 'negative' dimensions to territoriality: it can be both a force for oppression and also one for liberation. Particular strategies can be used to assert an identity and territorially transgressive acts can be employed to reclaim space and, hence, to assert basic rights.

Further reading

There is a wide range of useful social, cultural and urban geography texts dealing with many of the concerns of this chapter. Some of these are listed below along with work focusing specifically on issues of social and spatial marginalization and phenomena such as gentrification. Work highlighting the diverse contributions feminist geographic thought and practice have made to human geography and on the complex nature and meanings of home are also included.

Anderson, J. (2010) *Understanding Cultural Geography. Places and Traces*, London: Routledge.

Bagaeen, S. and Uduku, O.(eds) (2010) *Gated Communities. Social Sustainability and Historical Gated Developments*, London: Earthscan.

Blunt, A. and Dowling, R. (2006) *Home*, London: Routledge.

Browne, K., Lim, J. and Brown, G. (eds) (2007) *Geographies of Sexualities: Theory, Politics and Practice*, Aldershot: Ashgate.

Hubbard, P. (2006) *City*, London: Routledge.

Nelson, L. and Seager, J. (eds) (2004) *A Companion to Feminist Geography*, Oxford: Blackwell.

Panelli, R. (2004) *Social Geographies: From Difference to Action*, London: Sage.

Sibley, D. (1995) *Geographies of Exclusion: Society and Difference in the West*, London: Routledge.

Tucker, A. (2009) *Queer Visibilities: Space, Identity and Interaction in Cape Town*, Chichester: Wiley-Blackwell.

Valentine, G. (2001) *Social Geographies: Space and Society*, Harlow: Prentice Hall.

9 Conclusions

This book has explored various aspects of the construction of territories, the meanings attached to them and strategies associated with them. A range of examples has been utilized drawn from a variety of spatial scales ranging from the global to the local. These have contributed to a consideration of the ways in which territory is conceived, imagined and constructed. Attention has been drawn to the ways in which territory is often a major component of self-identity and, more significantly, group identity. Particular ideologies and social practices are manifested in space and territorial strategies are routinely employed to gain or to maintain power. A territorial frame is also regularly deployed in order to actively resist the imposition of power by dominant groups.

The significance of territorial thinking and the ways in which it reproduces divisions between social groups was emphasized. In broad terms, social practices are reflected in struggles and territorial claims over the use and control of space. Territorial strategies are utilized in conflicts concerned with social power and identity at a range of spatial scales from the global down to the very local. These strategies may be to do with maintaining power or with resisting the imposition of power by a dominant group. Exclusionary ideologies may be consolidated and reinforced through territorial practices, yet they can also be resisted through similar means.

Much attention was devoted to the most obvious form of territorial division: the state system. In considering the evolution and sustainment of this system, there was a particular focus on the territorial ideology of nationalism, a key building block in the formation of present-day states. While states can be seen as spatial containers, they are much more than this: they play a role in the formation of people's sense of both self and collective identity. The existence of discrete territorial units, and the boundaries between them, reflects a world where control over territory can be seen to signify power. The division of geographic space into territories, at whatever scale, represents the spatial expression of power. Notwithstanding claims of the emergence of a borderless world and the imminent death of the state, linked to globalizing processes, issues of control over territory are likely to continue to be of significance. The persistence of national allegiances and the increasing number of independent states suggest that the death of the state and the demise of the significance of borders is a long way off.

The nature and scope of state sovereignty (a somewhat idealized concept) has always been subject to contestation, modification and adaptation and seems likely to continue in a similar vein. Within a more globalized world, power is unequally distributed and some political–territorial formations exert significantly more power than others. Sovereignty, for some, is contingent on adhering to particular ground rules laid out by the more powerful. While these are often couched in terms of civil rights issues, in reality they are usually bound up with the advancement of the strategic interests (economic and political) of powerful elites. The creation and maintenance of military bases in other countries and the pressures exerted (both directly and indirectly) by major powers on more subordinate states further highlights the contingent nature of state sovereignty. A neo-liberal economic orthodoxy points towards the irrelevancy of borders, while simultaneously utilizing and reconfiguring them. This is reflected in the debates surrounding international migration. While capital flows relatively freely across international frontiers, people (depending on who they are, where they are from, and where they are trying to go) are faced with the intimidating paraphernalia associated with state borders.

While the system of territorial states is the most obvious form of territoriality, numerous more micro-scale and often more banal examples also occur. Apart from administrative sub-divisions of the state, there is a current emphasis on locality and place-based communities, building on and (re)producing local territorial identities. While there may be positive outcomes associated with some of these territorially defined communities, problems may emerge in specific contexts. Some people may be excluded, or at least attempts may be made to exclude them, from certain spaces on the basis of 'race', ethnicity, gender, sexual orientation, or simply because they belong to a category seen as 'other'. These various manifestations of territoriality can be seen as power expressed through a territorial frame but they also reflect particular ideologies which are then transposed onto space.

Throughout this book a key idea has been that territorial strategies are useful organizing devices and, more significantly, they are a means through which power is maintained or contested. From the state downwards, territories are constructed and used as means of social, economic and political control. It follows from this that territories cannot be seen as naturally occurring entities; rather, they are human creations and their construction and reproduction represent ways in which we conceive of the world around us. Territorial behaviour in humans is not an innate tendency; rather, it is a product of social, cultural, economic and political circumstances. While recognizing the constructed nature of territoriality and the functions it serves, the importance of territorial arrangements in people's everyday lives cannot be ignored. Just because territories are constructed, it does not mean they are not 'real' and a territorial frame serves to link people to place. It is obvious that people form attachments to place and these can have a huge impact on how people think and act. The significance of nationalism, an ideology based on attachment to a 'homeland', is a prime example. Place undoubtedly matters.

Territorial constructions are used to impose forms of control but they are also resisted, as in secessionist nationalism, opposition to the EU or disputes surrounding slum settlements. From the activities of revolutionary groups such as the Zapatistas in Chiapas in Mexico to 'reclaim the night' marches, control of territory is subject to constant struggle as attempts are made to control space or to wrest space from other groups. While many people do not necessarily freely choose their 'place', they may, nevertheless, identify with their immediate neighbourhood or locality. This sense of identity can in turn be converted into forms of action aimed at obtaining particular outcomes. The formation of community or residence groups reflects feelings of belonging or attachment to a particular place. It follows that notions of territory are connected with ideas of social power. The claiming of space is a political act whether it occurs in the 'public' or 'private' arena (and the categorization and demarcation of these areas is a key expression of territoriality).

All forms of human territorial behaviour reflect power relations. Territorial control and the contestations over geographic space reflect the workings of particular political processes. Simply observing the spatial outcomes tells us little about causative processes. As Sack (1986) observes, an emphasis on territoriality can tend to obscure the real actors. By focusing on the units (whether formalized and rigidly demarcated or whether informal and hazily bounded), attention is switched away from the protagonists and onto the end product of their activity. It is the processes giving rise to the patterns that are of ultimate importance in determining the welfare of those living within particular defined spaces. Both the territorial formations and the mechanisms underpinning the territorializations require investigation. Territorial behaviour can be either oppressive or liberating and can be used to assert or suppress an identity. Territories provide a material expression of the fusion of meaning, power and social space (Delaney 2009). In other words, power permeates society and this is manifested spatially as well as in many other ways.

Bibliography

Adams, G. (1995) *Free Ireland. Towards a Lasting Peace*. Dingle: Brandon.

Agnew, J. (1994) 'The territorial trap: the geographical assumptions of international relations theory', *Review of International Political Economy*, 1 (1): 53–80.

Agnew, J. (1995) 'Postscript: federalism in post-Cold War era', in G. Smith (ed.) *Federalism. The Multiethnic Challenge*, London: Longman, 294–302.

Agnew, J. (2002) *Geopolitics. Revisioning World Politics*, 2nd edn, London: Routledge.

Agnew, J., Mitchell, K. and Ó Tuathail, G. (eds) (2008) *A Companion to Political Geography*, Oxford: Blackwell.

Ali, M. (2003) *Brick Lane*, London: Doubleday.

Ali, T. (2002) *The Clash of Fundamentalisms. Crusades, Jihads and Modernity*, London: Verso.

Alland, A. (Jr) (1972) *The Human Imperative*, New York: Columbia University Press.

Anderson, B. (1991) *Imagined Communities. Reflections on the Origin and Spread of Nationalism*, London: Verso.

Anderson, J. (2008) 'Partition, consociation, border-crossing: Some lessons from the national conflict in Ireland/Northern Ireland', *Nations and Nationalism*, 14 (1): 85–104.

Anderson, J. (2010) *Understanding Cultural Geography. Places and Traces*, London: Routledge.

Anderson, M. (1996) *Frontiers. Territory and State Formation in the Modern World*, Cambridge: Polity Press.

Antonsich, M. (2009) 'On territory, the nation-state and the crisis of the hyphen', *Progress in Human Geography*, 33 (6): 789–806.

Ardrey, R. (1967) *The Territorial Imperative. A Personal Inquiry into the Animal Origins of Property and Nations*, London: Collins.

Ascherson, N. (1996) *Black Sea. The Birthplace of Civilization and Barbarism*, London: Vintage.

Atkinson, R. and Blandy, S. (2006) *Gated Communities*, London: Routledge.

August, O. (2000) *Along the Wall and Watchtowers. A Journey down Germany's Divide*, London: Flamingo.

Bachrach, P. and Baratz, M. (1962) 'Two faces of power', *American Political Science Review*, 56: 947–952.

Bagaeen, S. and Uduku, O. (eds) (2010) *Gated Communities. Social Sustainability and Historical Gated Developments*, London: Earthscan.

Barham, E. (2003) 'Translating terroir: the global challenge of French AOC labelling', *Journal of Rural Studies* 19: 127–138.

Batuman, B. (2010) 'The shape of the nation: visual production of nationalism through maps in Turkey', *Political Geography*, 29 (4): 220–234.

Bauman, Z. (1998) *Globalization. The Human Consequences*, New York: Columbia University Press.

Bell, J. and Staeheli, L. (2001), 'Discourses of diffusion and democratization', *Political Geography*, 20 (2): 175–195.

Bennett, N. (2006) The Role and Effectiveness of Parish Councils in Gloucestershire: Adapting to New Modes of Rural Community Governance, unpublished MPhil thesis, University of Worcester.

Berg, L.D. and Vuolteenaho, J. (eds) (2009) *Critical Toponymies. The Contested Politics of Place Naming*, Farnham: Ashgate.

Billig, M. (1995) *Banal Nationalism*, London: Sage.

Birch, J. (2007) 'The Lezghi', in B.A. Brower and B.R. Johnston (eds) *Disappearing Peoples? Indigenous Groups and Ethnic Minorities in South and Central Asia*, Walnut Creek, CA: Left Coast Press, 207–220.

Black, J. (1998) *Maps and Politics*, London: Reaktion.

Blacksell, M. (2006) *Political Geography*, London: Routledge.

Blakely, E.J. and Snyder, M.G. (1997) *Fortress America. Gated Communities in the United States*, Washington: The Brookings Institution.

Blakkisrud, H. and Kolstø, P. (2011) 'From secessionist conflict toward a functioning state: Processes of state- and nation-building in Transnistria', *Post-Soviet Affairs*, 27 (2): 178–210.

Blaut, J. (1987) *The National Question,* London: Zed Books.

Blunt, A. and Dowling, R. (2006) *Home*, London: Routledge.

Bogdanor, V. (1999) 'Devolution: decentralization or disintegration?', *The Political Quarterly*, 70 (2): 185–194.

Boyce, D.G. (1982) *Nationalism in Ireland*, London: Croom Helm.

Brenner, N. (1998) 'Between fixity and motion: Accumulation, territorial organization and the historical geography of spatial scales', *Environment and Planning D: Society and Space*, 16 (5): 459–481.

Brighenti, A.M. (2010a) 'On territorology. Towards a general science of territory', *Theory, Culture & Society*, 27 (1): 52–72.

Brighenti, A.M. (2010b) 'At the wall: Graffiti writers, urban territoriality, and the public domain', *Space and Culture*, 13 (3): 315–332.

Brower, B.A. and Johnston, B.R. (eds) (2007) *Disappearing Peoples? Indigenous Groups and Ethnic Minorities in South and Central Asia*, Walnut Creek, CA: Left Coast Press.

Browne, K. and Lim, J. (2010) 'Trans lives in the "gay capital of the UK"', *Gender, Place & Culture*, 17 (5): 615–633.

Browne, K., Lim, J. and Brown, G. (eds) (2007) *Geographies of Sexualities: Theory, Politics and Practice*, Aldershot: Ashgate.

Brunnbauer, U. (2005) 'Ancient nationhood and the struggle for statehood: Historiographic myths in the Republic of Macedonia', in P. Kolstø (ed.) *Myths and Boundaries in South-Eastern Europe*, London: Hurst & Company, 262–296.

Bryan, D. (2000) *Orange Parades. The Politics of Ritual, Tradition and Control*, London: Pluto Press.

Bunce, M. (1994) *The Countryside Ideal. Anglo-American Images of Landscape*, London: Routledge.

Bunnell, T., Muzaini, H. and Sidaway, J.D. (2006) 'Global city frontiers: Singapore's hinterland and the contested socio-political geographies of Bintan, Indonesia', *International Journal of Urban and Regional Research*, 30 (1): 3–22.

Butler, T. (2007) 'For gentrification?' *Environment and Planning A*, 39 (1): 162–181.

Butler, T. and Lees, L. (2006) 'Super-gentrification in Barnsbury, London: globalization and gentrifying global elites at the neighbourhood level', *Transactions of the Institute of British Geographers NS*, 31 (4): 467–487.

Cameron, A. (2007) 'Geographies of welfare and exclusion: reconstituting the "public"', *Progress in Human Geography*, 31 (4): 519–526.

Campbell, D. (1999) 'Apartheid cartography: the political anthropology and spatial effects of international diplomacy in Bosnia', *Political Geography*, 18 (4): 395–435.

Carmody, P. (2009) 'Cruciform sovereignty, matrix governance and the scramble for Africa's oil: Insights from Chad and Sudan', *Political Geography*, 28 (6): 353–361.

Carson, C. (1998) *The Star Factory*, London: Granta.

Carter, N. (2007) *The Politics of the Environment. Ideas, Activism, Policy*, 2nd edn, Cambridge: Cambridge University Press.

Castells, M. (1997) *The Power of Identity*, Malden: Blackwell.

Castles, S. and Miller, M.J. (2009) *The Age of Migration. International Population Movements in the Modern World*, 4th edn, Basingstoke: Palgrave Macmillan.

Chatterjee, P. (1993) *The Nation and its Fragments. Colonial and Post-colonial Histories*, Cambridge: Cambridge University Press.

Chomsky, N. (2006) *Failed States. The Abuse of Power and the Assault on Democracy*, London: Penguin.

Church, A. and Coles, T. (eds) (2007) *Tourism, Power and Space*, London: Routledge.

Cleaver, F. (2001) 'Institutions, agency and the limitations of participatory approaches to development', in B. Cooke and U. Kothari (eds) *Participation: The New Tyranny?* London: Zed Books, 36–55.

Cloke, P., Crang, P. and Goodwin, M. (eds) (2005) *Introducing Human Geographies*, 2nd edn, London: Arnold.

Cooke, B. and Kothari, U. (eds) (2001) *Participation: The New Tyranny?* London: Zed Books.

Cosgrove, D. and Daniels, S. (eds) (1988) *The Iconography of Landscape*, Cambridge: Cambridge University Press.

Cox, K. R. (1998) 'Spaces of dependence, spaces of engagement and the politics of scale, or: looking for local politics', *Political Geography*, 17 (1): 1–23.

Cox, K. (2002) *Political Geography: Territory, State and Society*, Oxford: Blackwell.

Cox, K.R., Low, M. and Robinson, J. (eds) (2008) *The Sage Handbook of Political Geography*, London: Sage.

Crampton, J.W. (2011) 'Cartographic calculations of territory', *Progress in Human Geography* 35 (1): 92–103

Creswell, T. (2004) *Place. A Short Introduction*, Malden: Blackwell.

Cronin, M., Murphy, W. and Rouse, P. (eds) (2009) *The Gaelic Athletic Association 1884–2009*, Dublin: Irish Academic Press.

Crowley, E. (2006) *Land Matters. Power Struggles in Rural Ireland*, Dublin: Lilliput Press.

Crump, J. R. (2004) 'Producing and enforcing the geography of hate: race, housing segregation, and housing-related hate crimes in the United States', in C. Flint (ed.) *Spaces of Hate. Geographies of Discrimination and Intolerance in the USA*, New York: Routledge, 227–244.

Cullingworth, B. and Nadin, V. (2006) *Town and Country Planning in the UK*, 14th edn, London: Routledge.

Dahlman C.T. (2009) 'Territory', in C. Gallaher, C.T. Dahlman, M. Gilmartin, A. Mountz and P. Shirlow (eds) *Key Concepts in Political Geography*, London: Sage, 77–86.

Dahlman, C.T. and Williams, T. (2010) 'Ethnic enclavization and state formation in Kosovo', *Geopolitics*, 15 (2): 406–430.

Dalby S. (2007) 'Regions, strategies, and empire in the global war on terror', *Geopolitics*, 12 (4): 586–606.

Daniels, P., Bradshaw, M., Shaw, D. and Sidaway, J. (eds) (2012) *An Introduction to Human Geography. Issues for the 21st Century*, 4th edn, Harlow: Pearson.

Daniels, S. (1993) *Fields of Vision. Landscape Imagery and National Identity in England and the United States*, Cambridge: Polity Press.

Davidson, B. (1992) *The Black Man's Burden. Africa and the Curse of the Nation-State*, Oxford: James Currey.

Davies, R.R. (2000) *The First English Empire. Power and Identities in the British Isles 1093–1343*, Oxford: Oxford University Press.

Davis, M. (1990) *City of Quartz*, London: Verso.

Dawkins, R. (1976) *The Selfish Gene*, Oxford: Oxford University Press.

Delaney, D. (2005) *Territory. A Short Introduction*, Malden: Blackwell.

Delaney, D. (2009) 'Territory and territoriality', in R. Kitchin and N. Thrift (eds) *International Encyclopedia of Human Geography,* Oxford: Elsevier.

Delanty, G. and Kumar, K. (eds) (2006) *The Sage Handbook of Nations and Nationalism*, London: Sage.

Demetriou, S. (2003) 'Rising from the ashes? The difficult (re)birth of the Georgian state, in J. Milliken (ed.) *State Failure, Collapse and Reconstruction*, Malden: Blackwell, 105–129.

Dicken, P. (2011) *Global Shift. Transforming the World Economy*, 6th edn, London: Sage.

Dijkink, G. (1996) *National Identity and Geopolitical Visions. Maps of Pride and Pain*, London: Routledge.

Dikeç, M. (2006) 'Two decades of French urban policy: from social development of neighbourhoods to the republican penal state', *Antipode*, 38 (1): 59–81.

Dikeç, M (2007) 'Revolting geographies: urban unrest in France', *Geography Compass*, 1 (5): 1190–1206.

Dixon, D.P. and Jones, J.P. (2006) 'Feminist geographies of difference, relation and construction', in S. Aitken and G. Valentine (eds) *Approaches to Human Geography*, London: Sage, 42–56.

Dixon, J. A. and Durrheim, K. (2000) 'Displacing place identity: a discursive approach to locating self and other', *British Journal of Social Psychology*, 39 (1): 27–44.

Djilas, A. (2003) 'Funeral oration for Yugoslavia: an imaginary dialogue with western friends', in D. Djokic (ed.) *Yugoslavism. Histories of a Failed Idea 1918–1992*, London: Hurst & Company, 317–333.

Dobson, A. (2006) *Green Political Thought*, 4th edn, London: Routledge.

Dwyer, C. (2005) 'Migrations and diasporas', in P. Cloke, P. Crang, M. Goodwin (eds) *Introducing Human Geographies*, 2nd edn, London: Hodder Arnold, 495–508.

Edwards, B., Goodwin, M., Pemberton, S. and Woods, M. (2000) *Partnership Working in Rural Regeneration*, Bristol: Policy Press.

Elden, S. (2005) 'Missing the point: globalization, deterritorialization and the space of the world', *Transactions of the Institute of British Geographers*, 30 (1): 8–19.

Elden, S. (2007a) 'Governmentality, calculation, territory', *Environment and Planning D: Society and Space*, 25 (3): 562–580.

Elden, S. (2007b) 'Terror and territory', *Antipode*, 39 (5): 821–845.

Elden, S. (2009) *Terror and Territory. The Spatial Extent of Sovereignty*, Minneapolis: University of Minnesota Press.

Elden, S. (2010) 'Land, terrain, territory', *Progress in Human Geography*, 34 (6): 799–817.

England, K. (2006) 'Producing feminist geographies: theory, methodologies and research strategies', in S. Aitken and G. Valentine (eds) *Approaches to Human Geography*, London: Sage, 286–297.

Escolar, D. (2001) '"Emerging" indigenous identities on the Argentine-Chilean border: Subjectivity and the crisis of sovereignty among the Andean population of the San Juan Province', *CIBR Working Papers in Border Studies*, 01–4.

Escolar, M. (2003) 'Exploration, cartography and the modernization of state power', in N. Brenner, B. Jessop, M. Jones and G. MacLeod (eds) *State/Space. A Reader*, Oxford: Blackwell, 29–52.

Etherington, J. (2010) 'Nationalism, territoriality and national territorial belonging', *Papers*, 95 (2): 321–339.

Euskirchen, M., Lebuhn, H. and Ray, G. (2007) 'From borderline to borderland. The changing European border regime', *Monthly Review*, 59 (6): 41–52.

Faulks, K. (2000) *Citizenship*, London: Routledge.

Fall, J.J. (2010) 'Artificial states? On the enduring geographical myth of natural borders', *Political Geography*, 29 (3): 140–156.

Fell, A. (1991) 'Penthesilea, perhaps', in M. Fisher and U. Owen (eds) *Whose Cities?* London: Penguin, 73–84.

Ferguson, J. (2005) 'Seeing like an oil company: space, security, and global capital in neoliberal Africa', *American Anthropologist*, 107 (3): 377–382.

Flint, C. and Taylor, P.J. (2007) *Political Geography. World-Economy, Nation-State and Locality*, 5th edn, Harlow: Prentice Hall.

Fonseca, I. (1996) *Bury me Standing*, London: Vintage.

Forrest, J.B. (1988) 'The quest for state hardness in Africa', *Comparative Politics*, 20: 423–441.

Foucault, M. (1980) *Power/Knowledge. Selected Interviews and Other Writings* (C. Gordon, ed.), Brighton: Harvester Press.

Friedman, T.L. (2007) *The World is Flat. A History of the Twenty First Century*, expanded edn, New York: Farrar, Straus and Giroux.

Fukuyama, F. (1992) *The End of History and The Last Man*, New York: Free Press.

Fyfe, N. (1991) 'The police, space and society: the geography of policing', *Progress in Human Geography* 15 (3): 249–267.

Gagnon, V.P. (2006) *The Myth of Ethnic War: Serbia and Croatia in the 1990s*, Ithaca, NY: Cornell University Press.

Gallagher, C., Dahlman, C.T., Gilmartin, M., Mountz, A. and Shirlow, P. (2009) *Key Concepts in Political Geography*, London: Sage.

Gellner, E. (1983) *Nations and Nationalism*, Oxford: Blackwell.

Gellner, E. (1994) *Encounters with Nationalism*, Oxford: Blackwell.

Gellner, E. (1997) *Nationalism*, London: Wiedenfield and Nicholson.

Gilmartin, M. (2008) 'Migration, identity and belonging', *Geography Compass*, 2 (6): 1837–1852.

Gilmartin, M. (2009) 'Border thinking: Rossport, Shell and the political geographies of a gas pipeline', *Political Geography*, 28 (5): 274–282.

Glassner, M.I (1993) *Political Geography*, New York: John Wiley and Sons.

Glasze, G., Webster, C. and Frantz, K. (2006) *Private Cities. Global and Local Perspectives*, London: Routledge.

Glenny, M. (1996) *The Fall of Yugoslavia. The Third Balkan War*, 3rd edn, Harmondsworth: Penguin.

Glenny, M. (1999) *The Balkans 1804–1999: Nationalism, War and the Great Powers*, London: Granta.

Gold, J.R. (1982) 'Territoriality and human spatial behaviour', *Progress in Human Geography*, 6 (1): 44–67.

Gottman, J. (1973) *The Significance of Territory*, Charlottesville: University Press of Virginia.

Gould, S.J. (1983) *The Mismeasure of Man*, London: Penguin.

Gould, S.J. (1991) *The Flamingo's Smile*, London: Penguin.

Gowan, T. (2010) *Hobos, Hustlers and Backsliders. Homeless in San Francisco*, Minneapolis: University of Minnesota Press.

Graham, B. (1994) 'No place of the mind: contested Protestant representations of Ulster', *Ecumene*, 1 (3): 257–281.

Graham, B. (ed.) (1997) *In Search of Ireland. A Cultural Geography*, London: Routledge.

Graham, B. (ed.) (1998) *Modern Europe. Place, Culture and Identity*, London: Arnold.

Graham, B., Ashworth, G. and Tunbridge, J.E. (2000) *A Geography of Heritage. Power, Culture and Economy*, London, Arnold.

Graham, B. and Howard, P. (eds) (2008) *The Ashgate Research Companion to Heritage and Identity*, Aldershot: Ashgate.

Gramsci, A. (1971) *Selections from the Prison Notebooks*, New York: International Publishers.

Grant, R. (2005) 'The emergence of gated communities in a West African context: Evidence from Greater Accra, Ghana', *Urban Geography*, 26 (8): 661–683.

Gregory, D. (2004) *The Colonial Present. Afghanistan, Palestine, Iraq*, Malden: Blackwell.

Gregory, D. (2007) 'Vanishing points. Law, violence and exception in the global war prison', in D. Gregory and A. Pred (eds) *Violent Geographies: Fear, Terror, and Political Violence*, London: Routledge, 205–236.

Gregory, D. and Pred, A. (eds) (2007) *Violent Geographies: Fear, Terror, and Political Violence*, London: Routledge.

Gruffud, P. (1995) 'Remaking Wales: Nation-building and the geographical imagination, 1925–50', *Political Geography*, 14 (3): 219–239.

Guibernau, M. (1995) 'Spain: a federation in the making?', in G. Smith (ed.) *Federalism. The Multiethnic Challenge*, London: Longman, 239–254.

Guibernau, M. (1996) *Nationalism. The Nation-State and Nationalism in the Twentieth Century*, Cambridge: Polity Press.

Guibernau, M. (2007) *The Identity of Nations*, Cambridge: Polity Press.

Gupta, S. and Omoniyi, T. (eds) (2007) *The Cultures of Economic Migration. International Perspectives*, Aldershot: Ashgate.

Hall, E.T. (1959) *The Silent Language*, Garden City: Doubleday.

Hanson, S. and Pratt, G. (1995) *Gender, Work and Space*, London: Routledge.

Hardt, M. and Negri, A. (2000) *Empire*, Cambridge: Harvard University Press.

Harley, J.B. (1988) 'Maps, knowledge and power', in D. Cosgrove and S. Daniels (eds) *The Iconography of Landscape*, Cambridge: Cambridge University Press, 277–312.

Harris, E. (2009) *Nationalism. Theories and Cases*, Edinburgh: Edinburgh University Press.

Hartshorne, R. (1950) 'The functional approach in political geography', *Annals of the Association of American Geographers*, 40 (1): 95–130.

Harvey, D. (1989) *The Condition of Postmodernity*, Oxford: Blackwell.

Harvey, D. (1990) 'Between space and time: reflections on the geographical imagination', *Annals of the Association of American Geographers*, 80 (3): 418–434.

Harvey, D. (1996) *Justice, Nature and the Geography of Difference*, Oxford: Blackwell.

Harvey, D. (2000) *Spaces of Hope*, Berkeley: University of California Press.

Harvey, D. (2003) *The New Imperialism*, Oxford: Oxford University Press.

Harvey, D. (2005) *A Brief History of Neoliberalism*, Oxford: Oxford University Press.

Harvey, D. (2006) *Spaces of Global Capitalism. Towards a Theory of Uneven Geographical Development*, London: Verso.

Harvey, D. (2010) *The Enigma of Capital and the Crises of Capitalism*, London: Profile Books.

Hastings J.V. (2009) 'Geographies of state failure and sophistication in maritime piracy hijackings', *Political Geography*, 28 (4): 213–223.

Hay, C. (1996) *Re-stating Social and Political Change*, Buckingham: Open University Press.

Hayter, T. (2000) *Open Borders. The Case against Immigration Controls*, London: Pluto.

Hechter, M. (1975) *Internal Colonialism: The Celtic Fringe in British National Development*, London: Routledge.

Held, D. (1989) *Political Theory and the Modern State*, Cambridge: Polity Press.

Held, D. (ed.) (2004) *A Globalizing World? Culture, Economics, Politics*, 2nd edn, London: Routledge.

Held, D. and McGrew, A. (eds) (2002) *Governing Globalization. Power, Authority and Global Governance*, Cambridge: Polity.

Held, D. and McGrew, A. (2007) *Globalization/AntiGlobalization. Beyond the Great Divide*, 2nd edn, Cambridge: Polity.

Held, D. and McGrew, A. (eds) (2007) *Globalization Theory. Approaches and Controversies*, Cambridge: Polity.

Held, D., McGrew, A., Goldblatt, D. and Perraton, J. (1999) *Global Transformations. Politics, Economics and Culture*, Cambridge: Polity Press.

Herb, G.H. (1999) 'National identity and territory', in G.H. Herb and D.H. Kaplan (eds) *Nested Identities: Nationalism, Territory and Scale*, Lanham: Rowman and Littlefield, 9–30.

Herod, A. and Wright, M.W. (eds) (2002) *Geographies of Power: Placing Scale*, Oxford: Blackwell.

Hewitt, R. (2010) *Map of a Nation. A Biography of the Ordnance Survey*, London: Granta.

Hinde, R.A. (1987) *Individuals, Relationships and Culture: Links between Ethology and the Social Sciences*, Cambridge: Cambridge University Press.

Hobsbawm, E. (1990) *Nations and Nationalism since 1780. Programme, Myth, Reality*, Cambridge: Cambridge University Press.

Hobsbawm, E. (1998) *On History*, London: Abacus.

Hobsbawm, E. and Ranger, T. (eds) (1992) *The Invention of Tradition*, Cambridge: Cambridge University Press.

Hoffman, J. (2004) *Citizenship Beyond the State*, London: Sage.

Holloway, L. and Hubbard, P. (2001) *People and Place. The Extraordinary Geographies of Everyday Life*, Harlow: Prentice Hall.

Holloway, S.L. (2005) 'Articulating otherness? White rural residents talk about Gypsy-Travellers', *Transactions of the Institute of British Geographers*, 30 (3): 351–367.

Holloway, S. and Valentine, G. (eds) (2000) *Children's Geographies. Playing, Living, Learning*, London: Routledge.

Holmes, M. and Storey, D. (2011) 'Transferring national allegiance: cultural affinity or flag of convenience?' *Sport in Society*, 14 (2): 253–271.

Hooson, D. (ed.) (1994) *Geography and National Identity*, Oxford: Blackwell.

Hroch, M. (1993) 'From national movement to the fully formed nation', *New Left Review*, 198: 3–20.

Hubbard, P. (2006) *City*, London: Routledge.

Hudson R. (2000) 'Identity and exclusion through language politics: the Croatian example', in R. Hudson and F. Reno (eds) *Politics of Identity. Migrants and Minorities in Multicultural States*, Basingstoke: Palgrave, 243–264.

Ilbery, B., Morris, C., Buller, H., Maye, D. and Kneafsey, M. (2005) 'Product, process and place. An examination of food marketing and labelling schemes in Europe and North America', *European Urban and Regional Studies*, 12 (2): 116–132.

Ingold, T. (1986) *The Appropriation of Nature. Essays on Human Ecology and Social Relations*, Manchester: Manchester University Press.

Ingram, A. and Dodds, K. (eds) (2009) *Spaces of Security and Insecurity: Geographies of the War on Terror*, Aldershot: Ashgate.

James, P.E. (1969) *Latin America*, 4th edn, New York: Odyssey Press.

Jarman, N. (1998) 'Painting landscapes: the place of murals in the symbolic construction of urban space', in A.D. Buckley (ed.) *Symbols in Northern Ireland*, Belfast: Institute of Irish Studies, Queen's University, 81–98.

Jeffrey, A. (2006) 'Building state capacity in post-conflict Bosnia and Herzegovina: the case of Brcko District', *Political Geography*, 25 (2): 203–227.

Jerndal, R. and Rigg, J. (1998) 'Making space in Laos: constructing a national identity in a "forgotten" country', *Political Geography*, 17 (7): 809–831.

Jessop. B. (2003) *The Future of the Capitalist State*, Oxford: Polity Press.

Johnson, N. (1995) 'Cast in stone: monuments, geography and nationalism', *Environment and Planning D: Society and Space*, 13 (1): 51–65.

Johnston, R.J. and Sidaway, J. (2004), *Geography and Geographers: Anglo-American Human Geography Since 1945*, 4th edn, London: Arnold.

Jones, M., Jones, R. and Woods, M. (2004) *An Introduction to Political Geography. Space, Place and Politics*, London: Routledge.

Jones, R. (2009) 'Sovereignty and statelessness in the border enclaves of India and Bangladesh', *Political Geography*, 28 (6): 373–381.

Jones, R. and Merriman, P. (2009) 'Hot, banal and everyday nationalism: bilingual road signs in Wales', *Political Geography*, 28 (3): 164–173.

Journal of Rural Studies (2009) Special Issue – De-centring White Ruralities: Ethnicity and Indigeneity, 25 (4).

Kabeer, N. (ed.) (2005) *Inclusive Citizenship. Meanings and Expressions*, London: Zed Books.

Kamenka, E. (1976) *Nationalism: The Nature and Evolution of an Idea*, London: Edward Arnold.

Kellas, J.G. (1991) *The Politics of Nationalism and Ethnicity*, Basingstoke: Macmillan.

Klein, N. (2000) *No Logo: No Space, No Choice, No Jobs*, London: Flamingo.

Klein, N. (2007) *The Shock Doctrine: The Rise of Disaster Capitalism*, London: Allen Lane.

Kleveman, L. (2003) *The New Great Game. Blood and Oil in Central Asia*, London: Atlantic Books.

Knox, P. and Pinch, S. (2000) *Urban Social Geography. An Introduction*, 4th edn, Harlow: Prentice Hall.

Knox, P., Agnew, J. and McCarthy, L. (2008) *The Geography of the World Economy*, 5th edn, London: Hodder.

Kobayashi, A. and Peake, L. (2000) 'Racism out of place: thoughts on whiteness and an antiracist geography in the new millennium', *Annals of the Association of American Geographers*, 90 (2): 392–403.

Kofman, E. (2008) 'Feminist transformations of political geography', in K.R. Cox, M. Low and J. Robinson (eds) *The Sage Handbook of Political Geography*, London: Sage, 73–86.

Kohn, H. (1967) *The Idea of Nationalism*, 2nd edn, New York: Collier-Macmillan.

Kolstø, P. (ed.) (2005) *Myths and Boundaries in South-Eastern Europe*, London: Hurst & Company.

Kolstø, P. (2005) 'Introduction: Assessing the role of historical myths in modern society', in P. Kolstø (ed.) *Myths and Boundaries in South-Eastern Europe*, London: Hurst & Company, 1–34.

Kreutzman, H. (2007) 'The Wakhi and Kirghiz of the Pamirian Knot', in B.A. Brower and B.R. Johnston (eds) *Disappearing Peoples? Indigenous Groups and Ethnic Minorities in South and Central Asia*, Walnut Creek, CA: Left Coast Press, 169–186.

Kumar, K. (2003) *The Making of English National Identity*, Cambridge: Cambridge University Press.

Landman, K. (2010) 'Gated minds, gated places: the impact and meaning of hard boundaries in South Africa', in S. Bagaeen and O. Uduku (eds) *Gated Communities. Social Sustainability and Historical Gated Developments*, London: Earthscan, 49–61.

Laurie, N., Dwyer, C., Holloway, S. and Smith, F. (1999) *Geographies of New Femininities*, Harlow: Longman.

Lefebvre, H. (1991) *The Production of Space*, Oxford: Basil Blackwell.

Le Goix, R. and Webster, C. (2008) 'Gated communities', *Geography Compass*, 2 (4): 1189–1214.

Lemanski, C. and Oldfield, S. (2009) 'The parallel claims of gated communities and land invasions in a southern city: Polarized state responses', *Environment and Planning A*, 41 (3): 634–648.

Ley, D. and Cybriwsky, R. (1974) 'Urban graffiti as territorial markers', *Annals of the Association of American Geographers*, 64 (4): 491–505.

Leyshon, M. (2008) 'The betweenness of being a rural youth: inclusive and exclusive lifestyles', *Social and Cultural Geography*, 9 (1): 1–26.

Liepins, R. (2000) 'Exploring rurality through "community": discourses, practices and spaces shaping Australian and New Zealand rural "communities"', *Journal of Rural Studies*, 16 (3): 325–341.

Little, J. (2002) 'Rural geography: rural gender identity and the performance of masculinity and femininity in the countryside', *Progress in Human Geography*, 26 (5): 665–670.

Livingstone, D.N. (1992) *The Geographical Tradition. Episodes in the History of a Contested Enterprise*, Oxford: Blackwell

Lorenz, K. (1966) *On Aggression*, New York: Harcourt, Brace and World.

Lowenthal, D. (1994) 'European and English landscapes as national symbols', in D. Hooson (ed.) *Geography and National Identity*, Oxford: Blackwell, 15–38.

Lowenthal, D. (1998) *The Heritage Crusade and the Spoils of History*, Cambridge: Cambridge University Press.

Lowndes, V. and Sullivan, H. (2004) 'Like a horse and carriage or a fish on a bicycle: How well do local partnerships and public participation go together?', *Local Government Studies*, 30 (1): 51–73.

Lynn, N.J. and Fryer, P. (1998) 'National-territorial change in the republics of the Russian North', *Political Geography*, 17 (5): 567–588.

McCaughan, M. (2008) *The Price of our Souls. Gas, Shell and Ireland*, Dublin: Afri.

McCullough, M.E. (2008) *Beyond Revenge. The Evolution of the Forgiveness Instinct*, San Francisco: Jossey-Bass.

McDowell, L. (1999) *Gender, Identity and Place. Understanding Feminist Geographies*, Cambridge: Polity Press.

McGuinness, M. (2000) 'Geography matters? Whiteness and contemporary Geography', *Area*, 32: 225–230.

MacLaughlin J. (1986) 'The political geography of "nation-building" and nationalism in social sciences: structural vs. dialectical accounts', *Political Geography Quarterly*, 5 (4): 299–329.

MacLaughlin, J. (2001) *Reimagining the Nation-State. The Contested Terrains of Nation-Building*, London: Pluto.

Malmberg, T. (1980) *Human Territoriality. Survey of Behavioural Territories in Man with Preliminary Analysis and Discussion of Meaning*, The Hague: Mouton.

Manent, P. (1997) 'Democracy without nations?', *Journal of Democracy*, 8 (2): 92–102.

Mann, M. (1984) 'The autonomous power of the state', *European Journal of Sociology*, 25 (2): 185–213.

Marshall, T.H. (1950) *Citizenship and Social Class*, Cambridge: Cambridge University Press.

Marx, K. and Engels, F. (1969) *Manifesto of the Communist Party*, Moscow: Progress Publishers.

Massey, D. (1994) *Space, Place and Gender*, Cambridge: Polity Press.

Mawby, R.I. and Yarwood, R. (eds) (2011) *Rural Policing and Policing the Rural. A Constable Countryside?* Farnham: Ashgate.

Maye, D., Holloway, L. and Kneafsey, M. (eds) (2007) *Alternative Food Geographies. Representation and Practice*, Oxford: Elsevier.

Mazower, M. (2001) *The Balkans. A Short History*, London: Phoenix.

Megoran, N. (2006) 'For ethnography in political geography: experiencing and re-imagining Ferghana Valley boundary closures', *Political Geography*, 26 (10): 622–640.

Meinhof, U.H. (ed.) (2002) *Living (with) Borders. Identity Discourses on East-West Borders in Europe*, Aldershot: Ashgate.

Miliband, R. (1969) *The State in Capitalist Society*, London: Quartet.

Miliband, R. (1991) 'Reflections on the crisis of the communist regimes', in R. Blackburn (ed.) *After the Fall. The Failure of Communism and the Future of Socialism*, London: Verso, 6–17.

Miller, D. (1997) *On Nationality*, Oxford: Clarendon Press.

Milliken, J. (ed.) (2003) *State Failure, Collapse and Reconstruction*, Malden: Blackwell.

Milliken, J. and Krause, K. (2003) 'State failure, state collapse, and state reconstruction: concepts, lessons and strategies', in J. Milliken (ed.) *State Failure, Collapse and Reconstruction*, Malden: Blackwell, 1–21.

Mitchell, D. (2005) 'The S.U.V. model of citizenship: floating bubbles, buffer zones, and the rise of the "purely atomic" individual', *Political Geography*, 24 (1): 77–100.

Monbiot, G. (2000) *Captive State. The Corporate Takeover of Britain*, Basingstoke: Macmillan.

Moran, W. (1993) 'The wine appellation as territory in France and California', *Annals of the Association of American Geographers*, 83 (4): 694–717.

Morley, D. (2000) *Home Territories. Media, Mobility and Identity*, London: Routledge.

Morris, D. (1973) *Manwatching. A Field Guide to Human Behaviour*, London: Jonathan Cape.

Morris, D. (1994) *The Naked Ape. A Zoologist's Study of the Human Animal*, London: Vintage.

Morton, H.V. (2000) *In Search of England*, London: Methuen.

Moseley, M. (2002) *Rural Development. Principles and Practice*, London: Sage.

Moseley, M. (ed.) (2003) *Local Partnerships for Rural Development: The European Experience*, Wallingford: CABI.

Mosse, D. (2001) '"People's knowledge", participation and patronage: Operations and representations in rural development', in B. Cooke and U. Kothari (eds) *Participation: The New Tyranny?* London: Zed Books, 16–35.

Mountz, A. (2009) 'Border', in C. Gallaher, C.T. Dahlman, M. Gilmartin, A. Mountz and P. Shirlow (eds) *Key Concepts in Political Geography*, London: Sage, 198–209.

Muir, R. (1997) *Political Geography. A New Introduction*, Basingstoke: Macmillan.

Muir, R. (1999) *Approaches to Landscape*, Basingstoke: Macmillan.

Murphy, A. (1993) 'Linguistic regionalism and the social construction of space in Belgium', *International Journal of the Sociology of Language*, 104 (1): 49–64.

Murray, W.E. (2006) *Geographies of Globalization*, London: Routledge.

Nairn, T. (1977) *The Break-up of Britain*, London: New Left Books.

Nairn, T. (1988) *The Enchanted Glass. Britain and its Monarchy*, London: Chandos.

Nairn, T. (1997) *Faces of Nationalism. Janus Revisited*, London: Verso.

Nash, C. (1993) '"Embodying the nation": the west of Ireland landscape and Irish identity', in B. O'Connor and M. Cronin (eds) *Tourism in Ireland. A Critical Analysis*, Cork: Cork University Press, 86–112.

Nash, C. (2009) 'Irish place names: post-colonial locations', in L.D. Berg and J. Vuolteenaho (eds) *Critical Toponymies. The Contested Politics of Place Naming*, Farnham: Ashgate, 137–152.

Nelson, L. and Seager, J. (2004) *A Companion to Feminist Geography*, Oxford: Blackwell.

Newman, D. (2002) 'The geopolitics of peacemaking in Israel-Palestine', *Political Geography*, 21 (5): 629–646.

Newman, D. and Paasi, A. (1998) 'Fences and neighbours in the postmodern world: boundary narratives in political geography', *Progress in Human Geography*, 22 (2): 186–207.

NicCraith, M. (1995) 'The symbolism of language in Northern Ireland', in U. Kockel (ed.) *Landscape, Heritage and Identity. Case Studies in Irish Ethnography*, Liverpool: Liverpool University Press, 11–46.

Nijman, J. (2008) 'Against the odds: slum rehabilitation in neoliberal Mumbai', *Cities*, 25: 73–85.

Nijman, J. (2009) 'A study of place in Mumbai's slums', *Tijdschrift voor Economische en Sociale Geografie*, 101 (1): 4–17.

Nisbet, J. (2007) *Sources of the River. Tracking David Thompson across Western North America*, Seattle: Sasquatch Books.

Nodia, G. (1994) 'Nationalism and democracy', in L. Diamond and M.F. Plattner (eds) *Nationalism, Ethnic Conflict and Democracy*, Baltimore and London: Johns Hopkins University Press, 3–22.

O'Loughlin, J., Kolossov, V. and Toal, G. (2011) 'Inside Abkhazia: Survey of attitudes in a *de facto* state', *Post-Soviet Affairs*, 27 (1): 1–36.

O'Reilly, C. (1998) 'The Irish language as symbol: visual representations of Irish in Northern Ireland', in A.D. Buckley (ed.) *Symbols in Northern Ireland*, Belfast: Institute of Irish Studies, Queen's University, 43–62.

Ó Tuathail, G. (1996) *Critical Geopolitics. The Politics of Writing Global Space*, London: Routledge.

Ó Tuathail, G., Dalby, S. and Routledge, P. (eds) (2006) *The Geopolitics Reader*, London: Routledge.

Ohmae, K. (1996) *The End of the Nation State. The Rise of Regional Economies*, London: HarperCollins.

Oliveira, J., Roca, Z. and Leitao, N. (2010) 'Territorial identity and development: from topophilia to terraphilia', *Land Use Policy*, 27: 801–814.

Onians, J. (ed.) (2004) *The Art Atlas*, London: Laurence King.

Osborne, S.P., Williamson, A. and Beattie, R. (2004) 'Community involvement in rural regeneration partnerships: Exploring the rural dimension', *Local Government Studies*, 30 (2): 156–181.

Oslender, U. (2007) 'Spaces of terror and fear on Columbia's Pacific coast. The armed conflict and forced displacement among black communities', in D. Gregory and A. Pred, A. (eds) *Violent Geographies: Fear, Terror, and Political Violence*, London: Routledge, 111–132.

Paasi, A. (1999) 'Boundaries as social practice and discourse: the Finnish-Russian border', *Regional Studies*, 33 (7): 669–680.

Paasi, A. (2008) 'Territory', in J. Agnew, K. Mitchell and G. Toal (eds) *A Companion to Political Geography*, Malden: Blackwell, 109–122.

Pain, R.H. (1997) 'Social geographies of women's fear of crime', *Transactions of the Institute of British Geographers NS*, 22 (2): 231–244.

Pain, R. (2010) 'The new geopolitics of fear', *Geography Compass*, 4 (3): 226–240.

Pain, R. and Smith, S.J. (eds) (2008) *Fear. Critical Geopolitics and Everyday Life*, Aldershot: Ashgate.

Painter, J. (2006) 'Prosaic geographies of stateness', *Political Geography*, 25 (7): 752–774.

Painter, J. (2010) Rethinking territory, *Antipode*, 42 (5), 1098–1118.

Painter, J. and Jeffrey, A. (2009) *Political Geography*, 2nd edn, London: Sage.

Panelli, R. (2004) *Social Geographies. From Difference to Action*, London: Sage.

Park, B.-G. (2005) 'Spatially selective liberalization and graduated sovereignty: politics of neo-liberalism and "special economic zones" in South Korea', *Political Geography*, 24 (7): 850–873.

Paxman, J. (1998) *The English. Portrait of a People*, London: Michael Joseph.

Peet, R. (1998) *Modern Geographical Thought*, Oxford: Blackwell.

Penrose, J. (2002) 'Nations, states and homelands: territory and territoriality in nationalist thought', *Nations and Nationalism*, 8 (3): 277–297.

Pepper, D. (1996) *Modern Environmentalism: An Introduction*, London: Routledge.

Perica, V. (2005) 'The sanctification of enmity: Churches and the construction of founding myths of Serbia and Croatia', in P. Kolstø (ed.) *Myths and Boundaries in South-Eastern Europe*, London: Hurst & Company, 130–157.

Phillips, M. (ed.) (2005) *Contested Worlds. An Introduction to Human Geography*, Aldershot: Ashgate.

Piaget, J. and Inhelder, B. (1967) *The Child's Conception of Space*, New York: Norton.

Plamenatz, J. (1976) 'Two types of nationalism', in E. Kamenka (ed.) *Nationalism: The Nature and Evolution of an Idea*, London: Edward Arnold, 22–36.

Political Geography (2006) Special Issue – Geographies as Sexual Citizenship, 25 (8).

Political Geography (2010) Special Issue – the State of Critical Geopolitics, 29 (5).

Poulantzas, N. (1969) 'The problem of the capitalist state', *New Left Review*, 58: 119–133.

Pratt, G. (2005) 'Masculinity-feminity', in P. Cloke, P. Crang and M. Goodwin (eds) *Introducing Human Geographies*, 2nd edn, London: Hodder Arnold, 91–103.

Pred, A. (2007) 'Situated ignorance and state terrorism', in D. Gregory and A. Pred (eds) *Violent Geographies: Fear, Terror, and Political Violence*, London: Routledge, 363–384.

Prescott, J.R.V. (1987) *Political Frontiers and Boundaries*, London: Unwin Hyman.

Radcliffe, S.A. (1998) 'Frontiers and popular nationhood: geographies of identity in the 1995 Ecuador-Peru border dispute', *Political Geography*, 17 (3): 273–293.

Raento, P. and Watson, C.J. (2000) 'Gernika, Guernica, Guernica? Contested meanings of a Basque place', *Political Geography*, 19: 707–736.

Ralphs, R., Medina, J. and Aldridge, J. (2009) 'Who needs enemies with friends like these? The importance of place for young people living in known gang areas', *Journal of Youth Studies*, 12 (5): 483–500.

Rao, A. (2007) Peripatetic peoples and lifestyles in South Asia', in B.A. Brower and B.R. Johnston (eds) *Disappearing Peoples? Indigenous Groups and Ethnic Minorities in South and Central Asia*, Walnut Creek, CA: Left Coast Press, 53–72.

Ray, C. (1998) 'Territory, structures and interpretation – two case studies of the European Union's LEADER I programme', *Journal of Rural Studies*, 14 (1): 79–87.

Reid-Henry, S.M. (2007) 'Exceptional sovereignty? Guantanamo Bay and the re-colonial present', *Antipode*, 39 (4): 627–648.

Ritzer, G. (2007) *The McDonaldization of Society 5*, London: Pine Forge.

Robinson, G.M. and Pobric, A. (2006) 'Nationalism and identity in post-Dayton Accords: Bosnia-Hercegovina', *Tijdschrift voor Economische en Sociale Geografie*, 97 (3): 237–252.

Roca, Z. and Oliveira-Roca, M.N. (2007) 'Affirmation of territorial identity: a development policy issue', *Land Use Policy*, 24: 434–442.

Rose, S., Kamin, L.J. and Lewontin, R.C. (1990) *Not in our Genes: Biology, Ideology and Human Nature*, Harmondsworth: Penguin.

Rosen, G. and Razin, E. (2009) 'The rise of gated communities in Israel: reflections on changing urban governance in a neo-liberal era', *Urban Studies*, 46 (8): 1702–1722.

Rousseau, J.J. (1973) *The Social Contract and Discourses*, London: Dent.

Ruane, J. and Todd, J. (1996) *The Dynamics of Conflict in Northern Ireland. Power, Conflict and Emancipation*, Cambridge: Cambridge University Press.

Rumley, D. and Minghi, J. (eds) (1991) *The Geography of Border Landscapes*, London: Routledge.

Sack, R. (1983) 'Human territoriality: a theory', *Annals of the Association of American Geographers*, 73 (1): 55–74.

Sack, R. (1986) *Human Territoriality: Its Theory and History*, Cambridge: Cambridge University Press.

Said, E.W. (1992) *The Question of Palestine*, new edn, London: Vintage.

Said, E.W. (1995) *Orientalism. Western Conceptions of the Orient*, new edn, London: Penguin.

Saldanha, A. (2011) 'The concept of race', *Geography*, 96 (1): 27–33.

Samers, M. (2010) *Migration*, London: Routledge.

Sassen, S. (2006) *Territory, Authority, Rights: From Medieval to Global Assemblages*, Princeton: Princeton University Press.

Sawyer, A. (2004) 'Violent conflicts and governance challenges in West Africa: the case of the Mano River basin area', *Journal of Modern African Studies*, 42 (3): 437–463.

Schwarzmantel, J. (1994) *The State in Contemporary Society. An Introduction*, Hemel Hempstead: Harvester Wheatsheaf.

Scott, J.C. (1998) *Seeing Like a State: How Certain Schemes to Improve the Human Condition Have Failed*, New Haven: Yale University Press.

Seton-Watson, H. (1977) *Nations and States. An Enquiry into the Origins of States and the Politics of Nationalism*, London: Methuen.

Sharma, A. and Gupta, A. (eds) (2006) *The Anthropology of the State. A Reader*, Malden: Blackwell.

Sharma, A. and Gupta, A. (2006) 'Introduction: rethinking theories of the state in an age of globalization', in A. Sharma and A. Gupta (eds) *The Anthropology of the State. A Reader*, Malden: Blackwell, 1–41.

Shaw, K. (2008) 'Gentrification: What it is, why it is, and what can be done about it', *Geography Compass*, 2 (5): 1697–1728.

Shirlow, P. and Murtagh, B. (2006) *Belfast. Segregation, Violence and the City*, London: Pluto Press.

Short, J.R. (1991) *Imagined Country. Society, Culture and Environment*, London: Routledge.

Shubin, S. (2011) '"Where can a Gypsy stop?" Rethinking mobility in Scotland', *Antipode*, 42 (5): 494–524.

Sibalis, M. (2004) 'Urban space and homosexuality: the example of the Marais, Paris "Gay Ghetto"', *Urban Studies*, 41 (9): 1739–1758.

Sibley, D. (1995) *Geographies of Exclusion. Society and Difference in the West*, London: Routledge.

Sidaway, J.D. (2007) 'Enclave space: a new metageography of development?' *Area*, 39 (3): 331–339.

Smith, A.D. (1986) *The Ethnic Origins of Nations*, Oxford: Blackwell.

Smith, A.D. (1991) *National Identity*, London: Penguin.

Smith, A.D. (1995) *Nations and Nationalism in a Global Era*, Cambridge: Polity Press.

Smith, A.D. (1998) *Nationalism and Modernism. A Critical Survey of Recent Theories of Nations and Nationalism*, London: Routledge.

Smith, A.D. (1999) *Myths and Memories of the Nation*, Oxford: Oxford University Press.

Smith, A.D. (2001) *Nationalism: Theory, Ideology, History*, Cambridge: Polity Press.

Smith, G. (1995) 'Federation, defederation and refederation: from the Soviet Union to Russian statehood,' in G. Smith (ed.) *Federalism. The Multiethnic Challenge*, London: Longman, 157–179.

Smith, M.J. (2009) *Power and the State*, Basingstoke: Palgrave Macmillan.

Smith, N. (1996) *The New Urban Frontier. Gentrification and the Revanchist City*, London: Routledge.

Smith, N. (2003) *American Empire. Roosevelt's Geographer and the Prelude to Globalization*, Berkeley: University of California Press.

Smith, S.J. (2005) 'Society-space', in P. Cloke, P. Crang and M. Goodwin (eds) *Introducing Human Geographies*, 2nd edn, London: Arnold, 18–33.

Smyth, W.J. (2006) *Map-making, Landscapes and Memory. A Geography of Colonial and Early Modern Ireland c. 1530–1750*, Cork: Cork University Press.

Soja, E.W. (1971) *The Political Organization of Space*, Washington: Association of American Geographers.

Soyinka, W. (1996) 'The national question in Africa: internal imperatives', *Development and Change*, 27: 279–300.

Squire, G. and Gill, A. (2011) '"It's not all Heartbeat you know": Policing domestic violence in rural areas', in R.I. Mawby and R. Yarwood (eds) *Rural Policing and Policing the Rural. A Constable Countryside?* Farnham: Ashgate, 159–167.

Squires, P (ed.) (2008) *ASBO Nation, The Criminalization of Nuisance in Contemporary Britain*, Bristol: Policy Press.

Spain, D. (1992) *Gendered Spaces*, Chapel Hill: University of North Carolina Press.

Staeheli, L.A. (2008) 'Citizenship and the problem of community', *Political Geography*, 27 (1): 5–21.

Staeheli, L.A. and Mitchell, D. (2006) 'USA's destiny? Regulating space and creating community in American shopping malls', *Urban Studies*, 43 (5–6): 977–992.

Storey, D. (2002) 'Territory and national identity: examples from the former Yugoslavia', *Geography*, 87 (2): 108–115.

Storey D (2003) *Citizen, State and Nation*, Sheffield: Geographical Association.

Storey, D. (2009) 'Political Geography', in R. Kitchin and N. Thrift (eds) *International Encyclopedia of Human Geography,* Volume 8, Oxford: Elsevier, 243–253.

Storey, D. (2010) 'Partnerships, people and place: Lauding the local in rural development', in G. Halseth, S. Markey and D. Bruce (eds) *The Next Rural Economies: Constructing Rural Place in Global Economies*, Wallingford: CABI, 155–165.

Sumartojo, R. (2004) 'Contesting place. Antigay and lesbian hate crime in Columbus, Ohio,' in C. Flint (ed.) *Spaces of Hate. Geographies of Discrimination and Intolerance in the USA*, New York: Routledge, 87–107.

Taylor, P.J. (1991) 'The English and their Englishness: "a curiously mysterious, elusive and little understood people"', *Scottish Geographical Magazine*, 107 (3): 146–161.

Taylor, P.J. (1994) 'The state as container: territoriality in the modern world-system', *Progress in Human Geography*, 18 (2): 151–162.

Thomson, S. (2005) '"Territorialising" the primary school playground: deconstructing the geography of playtime', *Children's Geographies*, 3 (1): 63–78.

Timothy, D.J. and Boyd, S.W. (2003*) Heritage Tourism*, London: Prentice Hall.

Toal, G. and Dahlman, C.T. (2011) *Bosnia Remade. Ethnic Cleansing and its Reversal*, Oxford: Oxford University Press.

Townshend, C. (2005) *Easter 1916: The Irish Rebellion*, London: Penguin.

Tuan, Y-F (1974) *Topophilia: A Study of Environmental Perception, Attitudes and Values*, Englewood Cliffs: Prentice-Hall.

Tucker, A. (2009) *Queer Visibilities: Space, Identity and Interaction in Cape Town*, Chichester: Wiley-Blackwell.

Ugarteche, O. (2000) *The False Dilemma. Globalization: Opportunity or Threat?* London/New York: Zed Books.

Unwin, T. (1991) *Wine and the Vine. An Historical Geography of Viticulture and the Wine Trade*, London: Routledge.

Urry, J. (2002) *The Tourist Gaze. Leisure and Travel in Contemporary Societies*, 2nd edn, London: Sage.

Valentine, G. (1989) 'The geography of women's fear', *Area*, 21 (4): 385–390.

Valentine, G. (1995) 'Out and about: geographies of lesbian landscapes', *International Journal of Urban and Regional Research*, 19 (1): 96–111.

Valentine, G. (2001) *Social Geographies. Space and Society*, Harlow: Prentice Hall.

Van Houtum, J. and Pijpers, R. (2007) 'The European Union as a gated community. The two-faced border and immigration regime of the EU', *Antipode*, 39 (2): 291–309.

Vasudevan, A., McFarlane, C and Jeffrey, A. (2008) 'Spaces of enclosure', *Geoforum*, 39 (5): 1641–1646.

Verbeek, P. (2008) 'Peace Ethology', *Behaviour*, 145 (11): 1497–1524.

Verdery, K. (1996) 'Whither "nation" and "nationalism"?', in G. Balakrishnan (ed.) *Mapping the Nation*, London: Verso, 226–234.

Vertovec, S. (2009) *Transnationalism*, London: Routledge.

Wallwork, J. and Dixon, J.A. (2004) 'Foxes, green fields and Britishness: On the rhetorical construction of place and national identity', *British Journal of Social Psychology*, 43 (1): 1–19.

Walters, W. (2004) 'The frontiers of the European Union: A geostrategic perspective', *Geopolitics*, 9: 674–698.

Warner, C.M. (1998) 'Sovereign states and their prey: The new institutionalist economics and state destruction in nineteenth-century West Africa', *Review of International Political Economy*, 5 (3): 508–533.

Weber, M. (1978) *Economy and Society. An Outline of Interpretive Sociology*, Berkeley: University of California Press.

Whatmore, S. (2002) *Hybrid Geographies. Natures, Cultures, Spaces*, London: Sage.

White, G.W. (1996) 'Place and its role in Serbian identity', in D. Hall and D. Danta (eds) *Reconstructing the Balkans. A Geography of the New Southeast Europe*, Chichester: John Wiley and Sons, 39–52.

White, G.W. (2000) *Nationalism and Territory. Constructing Group Identity in Southeastern Europe*, Lanham, MD: Rowman and Littlefield.

Wiener, M.J. (2004) *English Culture and the Decline of the Industrial Spirit, 1850–1980*, 2nd edn, Cambridge: Cambridge University Press.

Williams, C.H. and Smith, A.D. (1983) 'The national construction of social space', *Progress in Human Geography*, 7: 502–518.

Williams R. (1973) *The Country and the City*, London: Chatto and Windus

Winchester, H.P.M., Kong, L. and Dunn, K. (2003) *Landscapes. Ways of Imagining the World*, Harlow: Prentice Hall.

Winichakul, T. (1996) 'Siam mapped. The making of Thai nationhood', *The Ecologist*, 26 (5): 215–221.

Wood, D. (1992) *The Power of Maps*, London: Routledge.

Woods, M. (2005) *Rural Geography. Processes, Responses and Experiences in Rural Restructuring*, London: Sage.

Woods, M. (2007) 'Engaging the global countryside: globalization, hybridity and the reconstitution of rural place', *Progress in Human Geography*, 31 (4): 485–507.

Workman, L. and Reader, W. (2008) *Evolutionary Psychology. An Introduction*, Cambridge: Cambridge University Press.

Yarwood, R. (2005) 'Beyond the rural idyll: images, countryside change and geography', *Geography*, 90 (1): 19–32.

Yarwood, R. (2007) 'The geographies of policing', *Progress in Human Geography*, 31 (4): 447–465.

Yarwood, R. and Edwards, B. (1995) 'Voluntary action in rural areas: the case of Neighbourhood Watch', *Journal of Rural Studies*, 11 (4): 447–459.

Yiftachel, O. (2006) *Ethnocracy. Land and Identity Politics in Israel/Palestine*, Philadelphia: University of Pennsylvania Press.

Yiftachel, O. and Ghanem, A. (2005) 'Understanding ethnocratic regimes: the politics of seizing contested territories', *Political Geography*, 23 (4): 647–676.

Index